A Primer of Mathematical Writing

A Primer of Mathematical Writing

Being a Disquisition on Having Your Ideas Recorded, Typeset, Published, Read, and Appreciated

Steven G. Krantz

American Mathematical Society
Providence, Rhode Island

1991 *Mathematics Subject Classification.* Primary 00A35; Secondary 00A06, 00–01.

Library of Congress Cataloging-in-Publication Data
Krantz, Steven G. (Steven George), 1951–
   A primer of mathematical writing : being a disquisition on having your ideas recorded, typeset, published, read & appreciated / Steven G. Krantz.
      p. cm.
   Includes bibliographical references (p.   –   ) and index.
   ISBN 0-8218-0635-1 (alk. paper)
   1. Mathematics—Authorship.   I. Title.
QA42.K73   1996
808′.0665—dc20

96-45732
CIP

This book is dedicated to Paul Halmos.
For the example he has set for us all.

Table of Contents

Preface

The past fifty years have not seen as much emphasis on the quality of mathematical writing as perhaps one would wish. Because of competition for grants and other accolades, we hasten our work into print. An obituary for Hans Heilbronn (1908-1975) asserted that, after he wrote (by hand) a draft of a paper, he would put it on the shelf for one year. Then he would come back to it with fresh eyes, read it critically, and rewrite it. In effect, after a year's time, Heilbronn was reading his own work as though he were unfamiliar with it and had to understand each point from first principles. It is perhaps worth dwelling on this exercise to see what we might learn from it.

There is no feeling quite like that which comes after you have proved a good theorem, or solved a problem that you have worked on for a long time. Driven by the heat of passion, the words burst forth from your pen, the definitions get punched into shape, the proofs are built and bent and patched and shored up, and out goes that preprint to an appreciative audience. The whole paper sparkles—both the correct parts and the incorrect parts. A friend of mine, who solved a problem after working on it to the exclusion of all else for over fifteen years, used to rise up in the middle of the night just to caress his manuscript lovingly.

In circumstances like these, you find it virtually impossible to distance yourself from the material. Everything is emblazoned in your own mind and is crystal clear; you are unable to take the part of the uninitiated reader. You are torn between the desire (expeditiously) to record and validate your ideas, and the desire to communicate.

In today's competitive world, you probably do not feel that you have the luxury of setting a new paper aside for a year. The paper could be scooped; the subject could take a different direction and leave your great advance in the dust; the NSF might cancel your grant; the dean might not give you a raise; you might not be invited to speak at that big conference coming up.

Now let us look through the other end of the telescope. The harsh reality is this: If you prove the Riemann hypothesis, or the three-dimensional Poincaré conjecture, or Fermat's Last Theorem, then the world is willing to forgive you a lot. It would be nice if your paper were well written, for then more people could benefit from it more quickly. But—even if the paper is abysmally written—a handful of experts will be able to slug their way through it, they will teach it to others, perhaps more transparent proofs may come out, textbooks will eventually appear. Science is a process that tends to work itself out.

In fact most of us do not produce work that is at the high level just described—certainly not consistently so. If your work is not written in a clear fashion, so that the reader may quickly apprehend what the paper is about, what the main results are, and how the arguments proceed, then there is a considerable likelihood that he will set the paper aside before reading much of it. Your work will not have the impact that you had hoped or intended.

I am certainly not writing this book to advocate that you set aside each of your papers for a year, in the fashion of Heilbronn, and then rewrite it. Rather, I am asking you to consider the value of learning to write. Heilbronn had his techniques for sharpening up his prose. Each of us must learn his own.

I know many examples of mathematicians A and B, of roughly similar talent, with the property that A has enjoyed much greater success than B, and considerably more recognition for his ideas, because A wrote his work in an appealing and readable fashion and B did not. The A's and B's that I am thinking about are not at the Fields Medal level; Fields Medalists are exceptional in almost every respect, and probably do not need my advice. Instead, the examples of which I speak are several notches down from that august level, like most of us.

Even if you accept my thesis—that it is worthwhile to learn to write mathematics well—you may feel that fine writing is not an avocation that you wish to pursue. Fair enough: if you had wanted to become a writer, then probably that is what you would have done. But I submit that a reasonable alternative might be to spend an hour or two with this book, and perhaps another hour or two considering how its precepts apply to your own writing. The result, I hope, will be that you will be a more effective writer and will derive more enjoyment from the writing process.

As a scholar, or a scientist, you do not make widgets, nor do you grow wheat, nor do you perform brain surgery. In fact what you do is manipulate ideas and report on the results. Usually this report is in written form. What you write is often important, and can have real impact. Freshman composition teachers at Penn State like to tell their students of the engineers at Three Mile Island, who wrote to the governor of Pennsylvania three times to tell him that a nuclear disaster was in the making at their power plant. Their prose was so garbled that the poor governor could not determine *what* in the world they were talking about. The rest is history.

The very act of writing has, in the last twenty years, taken on a new shape and form. Whereas, years ago, it entailed sharpening a quill and buying a bottle of ink, nowadays most of us do not even own a quill knife. Instead we boot up the computer and create a document in some version of TEX. This being the case, I have decided to devote a (large) portion of this book to *techniques* of effective writing and another (much smaller) portion to the *instruments* of modern writing. This book is intended in large part for the novice mathematician. Fresh from graduate school, such a person must engage in the struggle of figuring out how to survive in the profession. The lucky budding mathematician will have gone to a graduate program that provided experience in technical writing and the use of hardware and software. If not, then perhaps the person is presently in a working environment that makes it easy to learn the technical aspects of writing. But I think that it is useful to have a reference for these matters. I intend, with this book, to provide one.

My credentials for writing this book are simple: I have written about one hundred articles and have written or edited about fifteen books. I have received a certain amount of praise for my work, and even a few prizes; and I have received plenty of criticism. Let me assure you that one of the most important attributes of a good writer is an ability to listen to criticism and to learn from it. Anyone finds it difficult to read criticism without becoming defensive; nobody wants to be excoriated. But even the most negative, uncharitable review can contain useful information. You profit not at all by becoming emotional; but if you can use the criticism to improve your work then you have trumped the critic.

This book is a rather personal tract, containing personal recommendations that reflect my own tastes. I have reason to believe that many others share these tastes, but not all do. There certainly are treatments of the art of mathematical writing that are more scholarly than this one—I note particularly the book [Hig] of Higham. He has careful discussions of how to select a dictionary or a thesaurus, careful catalogings of British usage versus American usage, a history of mathematical notation, clever exercises for developing skill with English syntax, tutorials on revision, and so forth. Higham's book is a real labor of love, and I recommend it highly. But there is no sense for me to duplicate Higham's efforts. Here I will discuss how to write, why to write, and when to write. However, this is not a scholarly tract, and it is not a text. The book is intended, rather, to be some friendly advice from a colleague. If an Assistant Professor or Instructor were to come to my office and ask for suggestions about writing, then I might reply "Let's go to lunch and talk about it." This book comprises what I would say over the course of several such meetings.

In this book I shall not give an exhaustive treatment of grammar, nor of *any particular aspect* of writing. When I do go into some considerable detail, it is usually on a topic not given extensive coverage elsewhere. Examples of such topics are **(i)** How to organize a paper, **(ii)** How to organize a book, **(iii)** How to write a letter of recommendation in a tenure case, **(iv)** How to write a referee's report, **(v)** How to write a book review, **(vi)** How to write a talk, **(vii)** How to write a grant proposal, **(viii)** How to write your Vita.

I have adopted the practice of labeling *incorrect* examples of grammar and usage with the symbol ✠. I do this so that examples of what is wrong will not be mistaken for examples of what is right.

I have benefited enormously from many friends and colleagues who were kind enough to read various drafts of the manuscript for this book. Their comments were insightful, and in many cases essential. In some instances they saved me from myself. I would like particularly to mention Lynn S. Apfel, Sheldon Axler, Don Babbitt, Harold Boas, Robert Burckel, Joe Christy, John P. D'Angelo, John Ewing, Gerald B. Folland, Len Gillman, Robert E. Greene, Paul Halmos, David Hoffman, Gary Jensen, Judy Kenney, Donald E. Knuth, Silvio Levy, Chris Mahan, John McCarthy, Jeff McNeal, Charles Neville, Richard Rochberg, Steven Weintraub, and Guido Weiss. George Piranian generously exercised his editing skills on my manuscript, and to good effect. I thank Randi Ruden for sharing with me her keen sense of language and her sharp wit; she showed no mercy, and spared no pains, in correcting my language and my logic. Josephine S. Krantz provided valuable moral support. I find it a privilege to be part of a community of scholars that is so generous with its ideas. Pat Morgan, Antoinette Schleyer, and Jennifer Sharp of the American Mathematical Society gave freely of their copy editing skills. Barbara Luszczynska, our mathematics librarian, also gave me help in tracking down sources. My work at MSRI was supported in part by NSF grant DMS-9022140.

It would be impossible for me to enumerate, or to thank properly, all the excellent mathematical writers from whose work I have learned. They have set the example, over and over again, and I am merely attempting to explain what they have taught us. Several other authors have addressed themselves to the task of explaining how to write mathematics, or how to execute scientific writing, or simply how to write. Some of their work is listed in the Bibliography. (See also [Hig] for a truly extensive enumeration of the literature.) The present book interprets some of the same issues from my own point of view, and filtered through my own sensibilities. I hope that it is a useful contribution.

S.G.K.
St. Louis, Missouri

Chapter 1
The Basics

Against the disease of writing one must take special precautions, since it is a dangerous and contagious disease.

Peter Abelard
Letter 8, Abelard to Heloise

Judge an artist not by the quality of what is framed and hanging on the walls, but by the quality of what's in the wastebasket.

Anon., quoted by Leslie Lamport

It matters not how strait the gate,
How charged with punishments the scroll,
I am the master of my fate;
I am the captain of my soul.

W. E. Henley

Your manuscript is both good and original; but the part that is good is not original, and the part that is original is not good.

Samuel Johnson

In America only the successful writer is important, in France all writers are important, in England no writer is important, and in Australia you have to explain what a writer is.

Geoffrey Cotterel

It may be true that people who are merely mathematicians have certain specific shortcomings; however, that is not the fault of mathematics, but is true of every exclusive occupation.

Carl Friedrich Gauss
letter to H. C. Schumacher [1845]

In fifty years nobody will have tenure but everyone will have a Ph.D.

V. Wickerhauser

1.1   What It Is All About

In order to write effectively and well, you must have something to say. This sounds trite, but it is the single most important fact about writing. In order to write effectively and well, you also must have an audience. And you must *know consciously* who that audience is. Much of the bad writing that exists is performed by the author of a research paper who thinks that all his readers are Henri Poincaré, or by the author of a textbook who does not seem to realize that his readers will be students.

Good writing requires a certain confidence. You must be confident that you have something to say, and that this something is worth saying. But you also must have the confidence to know that "My audience is X and I will write for X." Many a writer of a mathematical paper seems to be writing primarily to convince himself that his theorem is correct, but not as an effort to communicate. Such an author is embarrassed to explain anything, and hides behind the details. Many a textbook author seems to be embarrassed to speak to the student in language that the student will apprehend. Such an author instead finds himself making excuses to the instructor (who either will not read the book, or will flip through it impatiently and entirely miss the author's efforts).

Imagine penning a poem to your one true love, all the while thinking "What would my English teacher think?" or "What would my pastor think?" or "What would my mother think?" Have the courage of your convictions. Speak to that person or to those people whom you are genuinely trying to reach. Know what it is that you want to say and then say it.

For a mathematician, the most important writing is the writing of a research paper. You have proved a nice theorem, perhaps a great theorem. You certainly have something to say. You also know exactly who your audience is: other research mathematicians who are interested in your field of study. Thus two of the biggest problems for a writer are already solved. The issue that remains is how to say it. Remember that if you pen a love letter to yourself, then it will have both the good features and the bad features of such an exercise: it will exhibit both passion and fervor, but it will tend to exclude the rest of the world. What do these remarks mean in practice? In particular, they mean that as you write you must think of your reader—not yourself. You must consider his convenience, and his understanding—not your own.

In the Sputnik era, some years ago, when mathematics departments and journals were growing explosively and everyone was in a rush to publish, it was common to begin a paper by writing "Notation is as in my last paper." Today, by contrast, there are truly gifted mathematicians who write papers that look like a letter home to Mom: they just start to write, occasionally starting a new paragraph when the text spills over onto a new page, never formally stating a theorem or even a definition, never coming to any particular point. The contents are divine, if only the reader can screw up the courage to pry them loose.

These last two are not the sorts of papers that you would want to read, so why torment your readers like this? Much of the remainder of this book will discuss ways to write your work so that people *will* want to read it, and will enjoy it when they do.

1.2 Who Is My Audience?

If you are writing a diary, then it may be safe to say that your audience is just yourself. (Truthfully, even this may not be accurate, for you may have it in the back of your mind that—like Anne Frank's diary, or Samuel Pepys's diary—this piece of writing is something for the ages.) If you are writing a letter home to Mom, then your audience is Mom and, on a good day, perhaps Pop. If you are writing a calculus exam, then your audience consists of your students, and perhaps some of your colleagues (or your chairman, if the chair is in the habit of reviewing your teaching). If you are writing a tract on handle-body theory, then your audience is probably a well-defined group of fellow mathematicians. Know your audience!

Keep in mind a specific person—somebody actually in your acquaintance—to whom you might be writing. If you are writing to yourself or to Mom, this is easy. If you are instead writing to your peers in handle-body theory, then think of someone in particular—someone to whom you could be explaining your ideas. This technique is more than a facile artifice; it helps you to picture what questions might be asked, or what confusions might arise, or which details you might need to trot out and explain. It enables you to formulate the explanation of an idea, or the clarification of a difficult point.

I cannot repeat too often this fundamental dictum: have something to say and know what it is; know *why* you are saying it; finally, know to whom you are saying it, and keep that audience always in mind.

1.3   Writing and Thought

The ability to think clearly and the ability to write clearly are inextricably linked. If you cannot articulate a thought, formulate an argument, marshal data, assimilate ideas, organize a thesis, then you will not be an effective writer. By the same token, you can use your writing as a method of developing and honing your thoughts (see [Hig] for an insightful discussion of this concept).

We all know that one way to work out our thoughts is to engage in an animated discussion with someone whom we respect. But you can instead, à la Descartes, have that discussion with yourself. And a useful way to do so is by writing. When I want to work out my thoughts on some topic—teaching reform, or the funding of mathematics, or the directions that future research in several complex variables ought to take, or my new ideas about domains with noncompact automorphism group—I often find it useful to write a little essay on the subject. For writing forces me to express my ideas clearly and in the proper order, to fill in logical gaps, to sort out hypotheses from blind assumptions from conclusions, and to make my point forcefully and clearly. Sometimes I show the resulting essay to friends and colleagues, and sometimes I try to publish it. But, just as often, I file it away on my hard disc and forget it until I have future need to refer to it.

The writing of research level mathematics is a more formal process than that described in the last paragraph, but it can serve you just as well. When you write up your latest ideas for dissemination and publication, then you must finally face the music. That "obvious lemma" must now be treated; the case that you did not really want to consider must be dispatched. The ideas must be put in logical order and the chain of reasoning forged and fixed. It can be a real pleasure to craft your latest burst of creativity into a compelling flood of logic and calculation. In any event, this skill is one that you are obliged to master if you wish to see your work in print, and read by other people, and understood and appreciated.

Once you apprehend the principles just enunciated, writing ceases to be a dreary chore and instead turns into a constructive activity. It becomes a new challenge that you can aim to perfect—like your tennis backhand or your piano playing. If you are the sort of person who sits in front of the computer screen befuddled, frustrated, or even angry, and thinks "I know just what my thoughts are, but I cannot figure out how to say them" then something is wrong. Writing should *enable* you to express your thoughts, not hinder you. I hope that reading this book will help you to write, indeed will enable you to write, both effectively and well.

1.4 Say What You Mean; Mean What You Say

You have likely often heard, or perhaps uttered, a sentence like

> As a valued customer of XYZ Co., your call is very important to us.   ✠

Or perhaps

> To assist you better, please select one of the following from our menu: ✠

What is wrong with these sentences? The first suggests that "your call" is a valued customer. Clearly that is not what was intended. A more accurately formulated sentence would be

> You are a valued customer of XYZ Co., and your call is very important to us.

or perhaps

> Because you are a valued customer of XYZ Co., your call is very important to us.

In the second example, the phrase "To assist you better" is clearly intended to modify "we" (that is, it is "we" who wish to better assist you); therefore a stronger construction is

> So that we may assist you better, please select an item from our menu . . . .

or perhaps

> We can assist you more efficiently if you will make a selection from the following menu.

What is the point here? Is this just pompous nit-picking? Assuredly not. Mathematics cannot tolerate imprecision. The nature of mathematical *notation* is that it tends to rule out imprecision. But the *words* that connect our formulae are also important. In the two examples given above, we may easily discern what the speaker intended; but, in mathematics, if you formulate your thoughts incorrectly then your point may well be lost. Here are a few more examples of sentences that do not convey what their authors intended:

> Having spoken at hundreds of universities, the brontosaurus was a large green lizard.   ✠

(Amazingly, this sentence is a slight variant of one that was uttered by a distinguished scholar who is world famous for his careful use of prose.)

> As in our food, we strive to be creative with keeping the highest quality in mind, we have in our wine selections also.
>
>   ✠

(This sentence was taken from the menu of a rather good St. Louis restaurant.)

> To serve you better, please form a line.   ✠

(How many times have you heard this at your local retailer's, or at the bank?)

The message here is a simple one: Make sure that your subject matches your verb. Make sure that your referents actually refer to the person or thing that is intended. Make sure that your participles do not dangle. Make sure that your clauses cohere. *Read each sentence aloud.* Does each one make sense? Would you *say this in a conversation?* Would you understand it if someone else said it?

Use words carefully. A well-trained mathematician is not likely to use the word "continuous" to mean "measurable" nor "convex" to mean "one-connected". However we sometimes lapse into sloppiness when using ordinary prose. Treat your dictionary as a close friend: consult it frequently. As a consequence, do not use "enervate" to mean "invigorate" nor "fatuous" to mean "overweight" nor "provenance" to denote a geopolitical entity. When I am being underhanded, it is not because I am short of help.

In life, we receive many different streams of ideas simultaneously, and we parallel-process them in that greatest of all CPU's—the human brain. We absorb and process information and knowledge in a nonlinear fashion. But written discourse is linearly ordered. Word k proceeds directly after word $(k-1)$. The distinction between written language as a medium and the information flow that we commonly experience is one of the barriers between you and good writing. As you read this book (which purports to tell you how to write), you will see passages in which I say "now I will digress for a moment" or "here is an aside." (In other places I put sentences in parentheses or brackets; or I use a footnote.) These are junctures at which I could not fit the material being discussed into strictly logical order. You will have to learn to wrestle with similar problems in your own writing. One version of writer's block is a congenital inability to address this linear vs. nonlinear problem. In this situation, nothing succeeds like success. I recommend that, next time you encounter this difficulty, address it head on. After you have defeated it a few times (not without a struggle!), then you will be confident that you can handle it in the future.

I have discoursed on accurate use of language in the technical sense. Now let me remark on more global issues. As you write, you must think not only about whether your writing is correct and appropriate, but also about where your writing will go and what it will do when it gets there.

I have already admonished you to know when to start writing. Namely, you begin writing when you have something to say and you know clearly to whom you wish to say it. You also must know when to stop writing. Stop when you have said what you have to say. Say it clearly, say it completely, say it forcefully, say it without leap or lacuna, but then shut up. To prattle on and on is not to convince further.

And never doubt that language is a weapon. "Sticks and stones may break my bones but words will never hurt me" is perhaps the most foolish sentence ever uttered. You can inflict more damage, more permanently, with words than you can with any weapon. You can manipulate more minds, and more people, with words than with any other device. For example, when journalists in the 1960's referred to "self-styled radical leader Abbie Hoffman", they downgraded Hoffman in people's minds. They never referred to Spiro Agnew as a "self-styled [you fill in the blank]" or to Gordon Liddy as a "self-styled . . . ". This moniker was reserved for Abbie Hoffman—and sometimes for Jerry Rubin and Mario Savio—and one cannot help but surmise that it was for a reason. By the same token, newspapers frequently spoke in the 1960's of "outside agitators" visiting university campuses. They were never described as "colloquium speakers" or "expert political consultants".

When a policeman addresses you by

> Sir, may I see your driver's license? Did you notice that red light back there?

then he is sending out one sort of signal. (Namely, you are clearly a law-abiding citizen and he is just doing his job by pulling you over and perhaps giving you a ticket.) When instead a cop in the station house says

> OK, Billy. Why don't you spill your guts? You know that those other bums aren't going to do a thing to protect you. All they care about is saving their own skins. Jacko already confessed to the heist and told us that you held the gun, Billy. Now we need to hear it from you. Make it easy on

> yourself, Billy: play ball with us and we'll play ball with
> you.

then he is sending out a different sort of signal. (Namely, by using the first name—and not "William", but "Billy"—he is undercutting the addressee's dignity; he is treating the person like a wayward child. Further, the policeman is cutting off the individual from his peers, making him feel as though he is on his own. He is suggesting—albeit vaguely—that he may be willing to cut a deal.)

When you are a person of some accomplishment, and some clout, then your writing carries considerable responsibility. Your words may have great effect. You must weigh the words, and weigh their impact, carefully.

I am going to conclude this section with a brief homily. (I promise that there will be no additional homilies in the book; you may even ignore this one if you wish.) Nikolai Lenin said that the most effective way to bring down a society is to corrupt its language.[1] You need only look around you to perceive the truth of this statement. When language is corrupted, then people do not communicate effectively. When they do not communicate effectively, then they cannot cooperate. When they cannot cooperate, then the fabric of civilization begins to unravel.

Some of us use the word "bad" to mean "good". We use the phrase "let us keep our neighborhoods safe and clean" to mean "let us segregate our schools and arm every home"; we use the phrase "I am for gun control and freedom of choice" to mean "I'm a liberal and you're not." We say "account executive" when we mean "sales clerk" and "sanitation engineer" when we mean "garbage man". We use the words "interesting" to mean "foolish", "imaginative" to mean "irresponsible", and "naive" to mean "idiotic". These observations are not just idle cocktail party banter. They are in fact indicative of barriers between certain social groups.

It is just the same in mathematics. When we use the word "proof" to mean "guesses based on computer printouts" (see [Hor]), when we use "theoretical mathematics" to mean "speculative mathematics" (see [JQ]), when we use the phrase "Charles mathematicians" to belittle the practitioners of traditional and hard-won modes of reasoning that

---

[1]A similar statement is attributed to John Locke in *On Human Understanding*.

have been developed over many centuries (see [Man]), when we use the phrase "new mathematics" to mean "facile intuition" (see [PS], [Man]), then we are corrupting our subject. These are gross examples, but the same type of corruption occurs in the small when we write our work sloppily or not at all. It is the responsibility of today's scholars to develop, nurture, and record our subject for future generations. Good writing is of course not an end in itself; writing is instead the means for achieving the important goal of communicating and preserving mathematics.

1.5   Proofreading, Reading for Sound, Reading for Sense

Proofreading is an essential part of the writing process. And it is not a trivial one. You do not simply write the words and then quickly scan them to be sure that there are no gross errors. Paul Halmos [Hig] says that he never publishes a word before he has read it six times. Not all of us are that careful, but the spirit of his practice is correct:

- One proofreading should be to check *spelling* and simple *syntax* errors (software can help with the former, and even with the latter—see Section 6.4).

- One proofreading should be for *accuracy*.

- One proofreading should be for *organization* and for *logic*.

- One proofreading should be for *sense*, and for the flow of the ideas.

- One proofreading should be for *sound*.

The great English stage actor Laurence Olivier used to rehearse Shakespeare by striding across the countryside and delivering his lines to herds of bewildered cattle. Understandably, you may be disinclined to emulate this practice when developing your next paper on p-adic L functions—especially if you live in Brooklyn. However note this: all the best writers whom I know read their work aloud to themselves. Reading your words aloud *forces* you to make sense of what you have written, and to deliver it as a coherent whole. If you have never tried this technique, then your first experience with it will be a revelation. You will find that you quickly develop a new sensitivity for sound and sense in your writing. You will develop an "ear". You will learn instinctively what works and what does not.

Consider these simple examples. Suppose that the Hemingway novel *For Whom the Bell Tolls* were instead entitled "Who the Dingdong Rings For"; or that the Thornton Wilder play *Our Town* were called "My Turf". Even though the sense of the titles has not been changed appreciably, we see that the alternative titles eschew all the poetry and imagery that is present in the originals. "For Whom the Bell Tolls" evokes powerful emotions; the proffered alternative falls flat. The title *Our Town* suggests one value system, while "My Turf" brings to mind another. One fancies that, if *The Scarlet Letter* had had a less poetic title (how about "Bad Girls Finish Last"), then perhaps Hester Prynne would have garnered only an "$A-$", or maybe even an "Incomplete".

Mathematicians rarely have to wrestle with these poetic questions. But we need to choose names for mathematical objects; we need to formulate definitions. We need to describe and to explain. The word "continuous" is a perfect name for a certain type of function; the alternative terminology "nonhypererratic" would be much less useful. The phrase "the point x lies in a relative neighborhood of P" conveys a world of meaning in an elegant and memorable fashion. Not by accident has this terminology has become universal. You should strive for this type of precision and elegance in your own writing.

William Shakespeare said that "...a rose by any other name would smell as sweet." This statement is true, and an apt observation, in the context of the dilemma that faced Romeo and Juliet. But the name of a person, place, or thing can profoundly affect its future. There will never be a great romantic leading man of stage and screen who is named Eggs Benedict and there will never be a Fields Medalist or other eminent mathematician named Turkey Tetrazzini. The name of an object does not change its properties (consult Saul Kripke's New Theory of Reference for more on this thought), but it can change the way that the object is perceived by the world at large. Bear this notion in mind as you create terminology, formulate definitions, and give titles to your papers and other works.

Have you ever noticed that, when you are reading a menu or listening to an advertisement, it never fails that the food being described contains "fresh creamery butter" and "pure golden honey"? The marketing people never say "this grub contains butter and honey," for there is nothing appealing about the latter statement. But the first two evoke images of delicious food. As mathematicians, we are not in the position of hawking victuals. But we still must make choices to convey most effectively a given message, and the spirit of that message. We want to inform, and also to inspire. Consider the sentence

> The conjecture of Gauss (1830) is false.

Contrast this rather bald statement with

> The lemmas of Euler (1766) and the example of Abel (1827) led Gauss to conjecture (1830) that all semistable curves are modular. The conjecture was widely believed, and more than fifty papers were written by Jacobi, Dirichlet, and Galois in support of it. To everyone's surprise and dismay, a counterexample was produced by Frobenius in 1902. This counterexample opened many doors.

There is no denying that the second passage puts the entire matter in context, tells the reader who worked on the conjecture and why, and also how the matter was finally resolved. There is a tradition in written mathematics to conform to the terse. In your own writing, consider instead the advantages of telling the reader what is going on.

My advice is not to agonize over each word as you write a first draft. Just get the ideas down on the page. But *do* agonize a bit when you are editing and proofreading. A passage that reads

> This is a very important operator, that has very specific properties, culminating in a very significant theorem.   ✠

is all right as a first try, but does not work well in the long run. It overuses the word "very". It does not flow smoothly. It makes the writer sound dull witted. Consider instead

> This operator will be significant for our studies. Its spectral properties, together with the fact that it is smoothing of order one, will lead to our first fundamental theorem.

The second passage differs from the first in that it has *content*. It says something. It flows nicely, and makes the writer sound as though he has something worthwhile to offer.

An amusing piece of advice, taken from [KnLR, p. 102], is never to use "very" unless you would be comfortable using "damn" in its place.

A good, though not ironclad, rule of thumb is not to use the same word, nor even the same sound, in two consecutive sentences. Of course you may reuse the word "the", and the nouns that you are discussing will certainly be repeated; but, if possible, do not repeat descriptive words and do not place words that sound similar in close proximity.

Also be careful of alliteration. Vice President Spiro Agnew, with the help of speech writer William Safire, earned for himself a certain reputation by using phrases like "pampered prodigies", "pusillanimous pussyfooters", "vicars of vacillation", and "nattering nabobs of negativism". Whatever élan accrued to Agnew by way of this device is probably not something that you wish to cultivate for yourself. Lyndon Johnson led us into an escalated Vietnam war by deriding "nervous nellies". The alliterative device is often suitable for polemicism or poetry, but rarely so for mathematics. For example

> This semisimple, sesquilinear operator serves to show sometimes that subgroups of S are sequenced.   ✠

does not sound like mathematics. The typical reader probably will pause, reread the sentence several times, and wonder whether the writer is putting him on. Better is

> Observe that this operator is both semisimple and sesquilinear. These properties can lead to the conclusion that if G is a subgroup of S then G is sequenced.

Notice how simple syntactical tricks are used to break up the alliteration, and to good effect.

The last two points—not to repeat words or sounds, and to avoid intrusive alliteration—illustrate the principle of "sound and sense". If you read your work aloud as you edit and revise, then you will pick out offending passages quickly and easily. With practice, you also will learn how to repair them. The result will be clearer, more effective writing.

1.6   Compound Sentences, Passive Voice

It would be splendid if we could all write with the artistry of Flaubert, the elegance of Shakespeare, and the wisdom of Goethe. In mathematical writing, however, such an abundance of talent is neither necessary nor called for. In developing an intuitionistic ethics ([Moo]), for example, one presents the ideas as part of a ritualistic dance: there is a certain intellectual pageantry that comes with the territory. In mathematics, what is needed is a clear and orderly presentation of the ideas.

Mathematics is already, by its nature, logically complex and subtle. The sentences that link the mathematics are usually most effective when they are simple, declarative sentences. Compound sentences should be broken up into simple sentences. Avoid run-on sentences at all cost. Here are some examples:

Rather than saying

> As we let x become closer and closer to 0, then y tends ever closer to t_0.   ✠

instead say

> When x is close to 0 then y is close to t_0.

Of course mathematical notation allows us to write $\lim_{x \to 0} y = t_0$ instead of either of these; this abbreviated presentation will, in many contexts, be more desirable.

Rather than saying

> If g is positive, f is continuous, the domain of f is open, and we further invoke Lemma 2.3.6, then the set of points at which $f \cdot g$ is differentiable is a set of the second category, provided that the space of definition of f is metrizable and separable.   ✠

instead say

> Let X be a separable metric space. Let f be a continuous function that is defined on an open subset of X. Suppose that g is any positive function. Using Lemma 2.3.6, we see that the set of points at which $f \cdot g$ is differentiable is of second category.

An alternative formulation, even clearer, is this:

> Let X be a separable metric space. Let f be a continuous function that is defined on an open subset of X. Suppose that g is any positive function. Define
>
> $$S = \{x : f \cdot g \text{ is differentiable at } x\}.$$
>
> Then, by Lemma 2.3.6, S is of second category.

Note the use of the words "suppose" and "define" to break up the monotony of "let". Observe how the formal definition of the set S clarifies the slightly awkward construction in the penultimate version of our statement.

It is tempting, indeed it is a trap that we all fall into, to overuse a single word that means "hence" or "therefore". An experienced mathematical writer will have a clutch of words (such as "thus", "so", "it follows that", "as a result", and so on) to use instead. A paragraph in which every sentence begins with "therefore", or with "let", or "so" can be an agony to read. Have alternatives at your fingertips.

In general, you should avoid introducing unnecessary notation. Mary Ellen Rudin's famous statement

> Let X be a set. Call it Y.

is funny because it is so ludicrous. But this example is not far from the way we write when we are seduced by notation. A statement like

> Let X be a compact metric subspace of the space Y. If f is a continuous, $\mathbb{R}$-valued function on that space then it assumes both a maximum and a minimum value.   ✠

suffers from giving names to the metric space, its superspace, the function, and the target space, and then never using any of them. Slightly better is

> Let X be a compact metric space. If f is a continuous, real-valued function on X then f assumes both a maximum and a minimum value.

Better still is

> A continuous, real-valued function on a compact metric space assumes both a maximum value and a minimum value.

The last version of the statement uses no notation, yet conveys the message both succinctly and clearly.

Paul Halmos [Ste] asserts that mathematics should be written so that it reads like a conversation between two mathematicians who are on a walk in the woods. The implementation of this advice may require some effort. If what you have in mind is a huge commutative diagram, or the determinant of a big matrix whose entries are all functions, then you will likely be unsuccessful in conveying your thoughts orally. You must think in terms of how you, or another reasonable person, would *understand* such a complicated object. Of course such understanding is achieved in bits and pieces, and it is achieved conceptually. That is how you will communicate your ideas during a walk in the woods.

One corollary of the "walk in the woods" approach to writing is that you should write for a reader who is not necessarily sitting in a library, with all the necessary references at his fingertips. To be sure, most any reader will have to look up a few things. But if the reader must race to the stacks at every other sentence, then you are making his work too hard, and your paper is far too difficult to follow. Supply the necessary detail, and the proper heuristic, so that even if the reader is not sure of a notion he will be able temporarily to suspend his disbelief and move on.

Most authorities believe that writing in the passive voice is less effective than writing in the active voice. To write in the active voice is to identify the agent of the action, and to emphasize that agent (see [Dup] for a powerful discussion of active voice vs. passive voice). For example,

The manifold M is acted upon by the Lie group G as follows: ✝

is less direct, and requires more words, than

The Lie group G acts on the manifold M as follows:

Likewise, the statement

It follows that the set Z will have no element of the set Y lying in it. ✝

can be more clearly expressed as

Therefore no element of Y lies in Z.

Even better is

The sets Y and Z are disjoint.

or

Therefore $Y \cap Z = \emptyset$.

Notice that the last version of the statement used one word, while the first version used fifteen. Also, a mathematician much more readily apprehends $Y \cap Z = \emptyset$ than he does a string of verbiage. Finally, coming up with the succinct fourth formulation required not only restating the proposition, but also thinking about its meaning. The result was plainly worth the effort.

In spite of these examples, and my warnings against passive voice, I must admit that passive voice gives us certain latitude that we do not want to forfeit. If, in the first example, you have reason to stress the role of the manifold M over the Lie group G, then you may wish to use passive voice. In the second example, it is unclear how the use of passive voice could add a useful nuance to your thoughts. As usual, you must let sound work with sense to convey your message.

As I have already noted, no rule of writing is unbreakable. The active voice is usually more effective than the passive voice. If you run the software Grammatik® (which checks not only spelling, but syntax and English usage) on the Gettysburg address, then the output report is quite critical. (See Section 6.4 for a discussion of Grammatik® and similar software.) It cites the Gettysburg address repeatedly for use of passive voice, and for expressing ideas in a needlessly complicated fashion (columnists Mike Royko and William Safire, among others, have written witty pieces about Grammatik® and its treatment of several well-known speeches and documents; see for example [Saf]). But Lincoln had a good ear. If he had begun the speech with

Our ancestors founded this country 87 years ago.

then he would have certainly followed the dictum of using the active voice and using simple declarative sentences. However he would not have set the beautiful pace and tone that "Fourscore and seven years ago our fathers brought forth on this continent, a new nation, ..." invokes. He would have jumped too quickly into the rather difficult subject matter of his speech. (See [SW] for the provenance of these last ideas.)

As mathematicians, we rarely will be faced with a choice analogous to Lincoln's. But the principle illustrated here is one worth appreciating.

1.7   Technical Aspects of Writing a Paper

Even when your paper is in draft form, your name should be on it. A date is helpful as well. Number the pages. Write on one side of the paper only. Give the paper a working title.

Is all this just too compulsive? No.

First, you must always put your name on your work to identify it as your own. If it contains a good idea, then you do not want someone else to walk off with it. Because you tend to generate so many different drafts and versions of the things that you write, you should date your work. I have even known mathematicians who put a time of day on each draft. (Of course a computer puts a date and time stamp on each *computer file* automatically; here I am discussing hard copy or paper drafts.)

You should write your affiliation—even on the draft. If you are usually at Harvard, then write that down. If instead you are spending the year in Princeton, write that down. The draft could, at some point, be circulated. People need to know where to find you. With this notion in mind, include your current *e*-mail address.

If your writing is highly technical, and you are deeply involved in working out a complicated idea, then you do not want to burden yourself with not knowing in which order the pages go. Be sure to number them. The numbering system need not be "1 2 3 4 5...". It could be "1*A* 1*B* 1*C*..." or "1_{cov} 2_{cov} 3_{cov} ..." (to denote your subsection on the all-important covering lemma). In a rough draft, self-serving numbering systems like these can be marvelously useful. On the preprint that you intend to circulate, use a traditional sequential method for numbering the pages.

Take a few moments to think about the numbering of theorems, definitions, and so forth. This task is important both in writing a paper and in writing a book. Some authors number their theorems from 1 to n, their definitions from 1 to k, their lemmas from 1 to p, their corollaries from 1 to r—each item having its own numbering system. Do not laugh: this describes the default in LaTeX. As a reader, I find this method maddening; for the upshot is that I can never find anything. For instance, if I am on the page that contains Lemma 1.6, then that gives me no clue about where to find Theorem 1.5. If, instead, all displayed items are numbered in sequence—Theorem 1.2 followed by Corollary 1.3 followed by Definition 1.4, etc.—then I always know where I am.

Having decided on the logic of your numbering system, you also need to decide how much information you want each number to contain. What does this mean? My favorite numbering system (in writing a book) is to let "$\langle\langle Item \rangle\rangle$ 3.6.4" denote the fourth displayed item (theorem or corollary or lemma or definition) in the sixth section of Chapter 3. If there is a labeled, displayed equation in the statement of the $\langle\langle Item \rangle\rangle$ then I label it (3.6.4.1). The good feature of this system is that the reader always knows precisely where he is, and can find anything easily. The bad feature is that the numbering system is a bit cumbersome. Other authors prefer to number displayed items within each section. Thus, in Section 6 of Chapter 3 the displayed items are numbered simply 1, 2, 3, ... When reference is later made to a theorem, the reference is phrased as "by Theorem 4 in Section 6 of Chapter 3" or "by Theorem 4 of Section 3.6." As you can see, this ostensibly simpler numbering system is cumbersome in its own fashion.

The main point is that you want to choose a numbering system that suits your purposes, and to use it consistently. You want to make your book or paper as easy as possible for your reader to study. Achieving this end requires that you attend to many small details. Your numbering system is one of the most important of these.

A final point is this: do not number every single thing in your manuscript. This dictum applies whether you are writing a paper or a book. I have seen mathematical writing in which every single paragraph is numbered. Such a device certainly makes navigation easy. But it is cumbersome beyond belief. Likewise do not number all formulas. You will only be referring to some of them, and the reader knows that. To number all formulas will create confusion in the reader's mind; he will no longer be able to discern what is important and what is less so.

Write on one side of the paper only. If you do not, and if you are writing something fairly technical and complicated (like mathematics), then you can become hopelessly confused when trying to find your place. In addition, you must frequently set two pages side by side—for the sake of comparing formulas, for instance. This move is easy with a manuscript written on one side, and nearly impossible with one that is not.

If you are scrupulous about not wasting paper, and insist on using both sides, then my advice is this: write drafts of your mathematical papers on one side of fresh paper. When that work is typed up and out the door, boldly X- out the writing on the front side of each page of your old drafts. Turn the paper over, and use it as scratch paper, or for your laundry list.

Of course printers print on one side, and the advice of the preceding two paragraphs is not relevant to the printing of a TEXed document. But most of us still do the initial composition of a mathematics paper by hand on lined paper. Consider the foregoing advice when doing so.

I suggest writing in ink. Pencil can smear, erasing can tear the page, and it is difficult to read a palimpsest. Also pencil-written material does not photocopy well. Blue pens do not photocopy well either. I always write with a black pen on either white or yellow paper. I write either with a fountain pen or a rolling writer or a fiber-tip pen so that the pen strokes are *dense* and *sharp* and *dark*. I write with a pen that does not skip or blot. If it begins to do either, I immediately discard it and grab a new one.

Of course you cannot erase words that are written with a pen; but you can cross them out, and that is much cleaner. It is easier to read a page written in bold black ink, and which includes some crossed out passages, than to decipher a page of chicken scratch layered over erased smears written with a pencil or written with a pen that is not working properly.

Be sure that your desk is well stocked with paper, pens, Wite-Out®, Post-it® notes, a stapler, staples, a staple remover, cellophane tape, paper clips, manila folders, manila envelopes, scissors, a dictionary, and anything else you may need for writing. Have them all at your fingertips. You do not want to interrupt the precious writing process by running around and looking for something trivial.

Do not write much on each page. I advise writing *large*, and double or triple spaced. The reason? First, you want to be able to insert passages, make editorial remarks, make corrections, and so forth. Second, a page full of cramped writing on every line is hard to read. Third, you can more easily rearrange material if there is just a little on each page. For example, if one page contains the statement of the main theorem and nothing else, another contains key definitions and nothing else, and so forth, then you can easily change the location of the main theorem in the body of the paper. If the main theorem is buried in a page with a great deal of other material, then moving it would involve either copying, or photocopying, or cutting with scissors.

Do not hesitate to use colored pens. For instance, you could be writing text in black ink, making remarks and notes to yourself (like "find this reference" or "fill in this gap") in red ink, and marking unusual characters in green ink. This may sound compulsive, but it makes the editing process much easier.

A good bibliography is an important component of scholarly work (more on bibliographies can be found in Sections 2.6, 5.5). Suppose that you are writing a paper with a modest number of references (about 25, say), and you are assigning an acronym to each one. For instance, [GH] could refer to the famous book by Griffiths and Harris. When you refer to this work while you are writing, use the acronym. Keep a sheet of notes to remind yourself what each acronym denotes. Do not worry about looking up the detailed bibliographic reference while you are engaged in writing; instead, compartmentalize the procedure. When you are finished writing the paper, you will have a complete, *informal* list of all your references. At that time, go to the library and find all the details. You will be able to do so in an hour or two of fairly relaxed work. Running back and forth to the library during the heat of writing passion is both counterproductive and a waste of time.

If you are adept at using your computer, then you can cut through a lot of the tedium of assembling bibliographies. See the discussion in Sections 2.6 and 5.5.

Let me make a general remark about the writing process. As you are writing a paper, there will be several junctures at which you feel that you need to look something up: either you cannot remember a theorem, or you have lost a formula, or you need to imitate someone else's proof. My advice is *not* to interrupt yourself while you are writing. Take your red pen and make a note to yourself about what is needed. But *keep writing.* When you are in the mood to write, you should take advantage of the moment and do just that. Interrupting yourself to run to the library, or for any other reason, is a mistake.

Write on a desk that is free of clutter. It is romantic, to be sure, to watch a film in which the writer labors furiously on a desk that is awash with papers, books, hamburger bags, ice cream containers, old coffee cups, last week's underwear, and who knows what else. Leave that stuff to the movies. Instead imagine tearing into page 33 of your manuscript and accidentally spilling a week-old cup of coffee and a piece of pepperoni pizza all over your project. Think of the time lost in mopping up the mess, separating the pages, trying to read what you wrote, copying your pages, and so forth. Enough said.

If you are going to drink coffee or a soda or eat a sandwich while you work, I suggest having the food on a small separate side table. This little convenience will force you to be careful, and if you do have an accident then it will not make a mess of your work.

Write in a place where you can concentrate without interruption. Whether you have music going, or a white noise machine playing, or a strobe light flashing is your decision. But if you are going to concentrate on your mathematics, it may take up to an hour to get the wheels turning, to fill your head with all the ideas you need, and to start formulating the necessary assertions. After you have invested the necessary time to tool up, you want to use it effectively. Therefore you do not want to be interrupted. Close the door and unplug the telephone if you must. Victor Hugo used to remove all his clothes and have his servant lock him in a room with nothing but paper and a pen. More-over, the servant guarded the door so that the great man would not be interrupted by so much as a knock. This method is not very practical, and is perhaps not well suited to modern living, but it is definitely in the right spirit.

1.8   More Specifics of Mathematical Writing

For the most part, the writing of mathematics is like the writing of English prose. Indeed, it *is* a part of the writing of English. (*Caveat:* I hope that my remarks have some universality, and apply even if you are writing mathematics in Tagalog or Coptic or Tlingit.) If you read your work aloud (I advocate this practice in Section 1.5), then you should be reading complete sentences that flow from one to the next, just as they do in good prose.

It is all too easy to write a passage like

Look at this here equation:

$$x^n + y^n = z^n. \quad \maltese$$

Much smoother is the passage

> The equation
> $$x^n + y^n = z^n$$
> tells us that Fermat's Last Theorem is still alive.

Another example of good sentence structure is

> Since
> $$A < B$$
> we know that . . . .

Notice that the the sentence reads well aloud: "Since A is less than B we know that . . .".

An aspect of writing that is peculiar to mathematics is the use of notation. Without good notation, many mathematical ideas would be difficult to express. With it, our writing has the potential to be elegant and compelling. A common misuse of notation is to put it at the beginning of a sentence or a clause. For example,

> Let f be a function. f is said to be *semicontinuous* if . . .
>   ✠

and

> For most points x, $x \in S$.   ✠

Even in these two simple examples you can begin to apprehend the problem: the eye balks at a sentence or clause that is begun with a symbol. You find yourself rereading the passage a couple of times in order to discern the correct sense. Much better is:

> A function f is said to be *semicontinuous* if . . .

and

> We see that $x \in S$ for most points x.

Observe that both of these revisions are easily comprehended the first time through. That is one of the goals of good writing.

Mathematical notation is often so elegant and compelling that we are tempted to overuse, or misuse, it. For example, the notation in the sentence "If $x > 0$ then $x^2 > 0$" is no hindrance, is easy to read, and tends to make the sentence short and sweet (nonetheless, there are those who would tender cogent arguments for "If a number is positive then so is its square."). By contrast, the phrase

> Every real, nonsquare $x < 0 \ldots$ ✠

is objectionable. The reason is that it is not clear, on a first reading, what is meant. Are you saying that "Every real, nonsquare x is negative" or are you saying "Every real, nonsquare x that is less than zero has the additional property $\ldots$". By strictest rules, the notation $<$ is a *binary connective*. The notation is designed for expressing the thought $A < B$. If that is not the exact phrase that fits into your sentence, then you had best not use this notation.

When you are planning a paper, or a book, you should try to plan your notation in advance. You want to be consistent throughout the work in question. To be sure, we have all seen works that, in Section 9, say "For convenience we now change notation." All of a sudden, the author stops using the letter H to denote a subgroup and instead begins to use H to denote a biholomorphic mapping. Amazingly, this abrupt device actually works much of the time. But you should avoid it. If you can, use the same notation for a domain in Section 10 (or Chapter 10) of your work that you used in Section 1 (or Chapter 1). Try to avoid local contradictions—like suddenly shifting your free variable from x to y. Try not to use the same character for two different purposes.

The last stipulation is not always easy to follow. Many of us commonly use i for the index of a series or sequence: $\sum_{i=1}^{\infty} a_i$ and $\{x_i\}_{i=1}^{\infty}$. No problem so far, but suppose that you are a complex analyst, and use i to denote a square root of -1. And now suppose that this last i occurs in some of your sequences and series. You can see the difficulties that would arise. A little planning can help with this problem, though in the end it may involve a great deal of tedious work to weed out all notational ambiguities.

Many a budding mathematician is seduced by mathematical notation. There was a stage in my education when I thought that all of mathematics should be written without words. I wrote long, convoluted streams of $\forall$, $\exists$, $\ni:$, $\Rightarrow$, $\equiv$, and so forth. This style would have served me well had I been invited to coauthor a new edition of *Principia Mathematica* (see [WR]). In modern mathematics, however, you should endeavor to use English—and to minimize the use of cumbersome notation. Why burden the reader with

$$\forall x \exists y, x \geq 0 \Rightarrow y^2 = x \qquad \maltese$$

when you can instead say

Every nonnegative real number has a square root.

The most important logical syllogism for the mathematician is *modus ponendo ponens*, or "if ... then". If you begin a sentence with the word "if", then do not forget to include the word "then". Consider this example:

If $x > 4$, $y < 2$, the circle has radius at least 6, the sky is blue, the circle can be squared. $\maltese$

Which part of this sentence is the hypothesis and which the conclusion? After a few readings you may be able to figure it out. If it were sensible mathematics then the mathematical meaning would probably give you some clues. But it is clearer to write

If $x > 4$, $y < 2$, the circle has radius at least 6, and the sky is blue, then the circle can be squared.

Following the dictum that shorter sentences are frequently preferable to longer ones, you can express the preceding thought even more succinctly as

Suppose that $x > 4$, $y < 2$, the circle has radius at least 6, and the sky is blue. Then the circle can be squared.

The word "then" is pivotal to the logical structure here. It acts both as a connective and as a sign post. The reader can (usually) figure out what is meant if the word "then" is omitted. But the reader should not *have* to do so. Your job as the writer is to perform this task *for* the reader. Mathematicians have a tendency to want to jam everything into one sentence. However, as the last example illustrates, greater clarity can often be achieved by breaking things up; this device also forces you to think more clearly and to organize your thoughts more effectively.

Mathematicians commonly write "If f is a continuous function, then prove X." A moment's thought shows that this is not the intended meaning: the desire to prove X is not contingent on the continuity of f. What is intended is "Prove that if f is a continuous function then X." In other words, the hypothesis about f is part of what needs to be proved.

Sometimes you need to write a sentence that treats a word as an object. Here is an example:

> We call Γ the *fundamental solution* for the partial differential operator L. We use the definite article "the" because, suitably normalized, there is only one fundamental solution.

I have oversimplified the mathematics here to make a typographical point. First, when you define a term (for the first time), you should italicize the word or phrase. Second, when you refer to a word (in this case "the") as the object of discussion, then put that word in quotation marks. For a variety of psychological reasons, writers often do not follow this rule. It is helpful to recall W. V. O. Quine's admonition: " 'Boston' has six letters. However Boston has 6 million people and no letters."

The phrase "if and only if" is a useful mathematical device. It indicates logical equivalence of the two phrases that it connects. While the phrase is surely used in some other disciplines, it plays a special role in mathematical writing; we should take some care to treat it with deference. Some people choose to write it as "if, and only if,"—with two commas. That is perfectly grammatical, if a little stilted. One habit that is unacceptable (because it sounds artificial and is difficult to read) is to begin a sentence with this phrase. For instance,

> If and only if x is nonnegative, can we be sure that the real
> number x has a real square root.   ✠

That is a painful sentence to read, whether the reading is done aloud
or *sotto voce*. Better is

> A real number x has a real square root if and only if $x \geq 0$.

An alternative form, not with universal appeal (but better than
beginning a sentence with "if and only if") is

> Nonnegative real numbers, and only those, have real square
> roots.

The neologism "iff", reputed to have been popularized by Paul Halmos, is a generally accepted abbreviation for "if and only if". This
is a useful bridge between the formality of "if and only if" and the
convenience of "if".

Word order can have a serious, if subtle, effect on the meaning (or
at least the nuance) of a sentence. The examples

> Yellow is the color of my true love's hair.

> My true love's hair has the color yellow.

> The hair, which is yellow, of my true love ...

each say something different, as they emphasize a different aspect—
either the color, or the person, or the hair—that is being considered.
(As an exercise, insert the word "only" into all possible positions in the
sentence

> I helped Carl prove quadratic reciprocity last week.

and watch the meaning change.)

In mathematics, word order can seriously alter the meaning of a sentence, with the result that the sentence is not immediately understood—
if at all. When you proofread your own work, you tend to supply meaning that is not actually present in the writing; the result is that you
can easily miss obscurity imposed by word order. Reading your work
aloud can help cut through the problem.

Do not overuse commas. I become physically ill when I see a sentence like

> We went to the store, to buy some potatoes.   ✠

Slightly more subtle, but still irksome, is

> Now that we have our hypotheses in place, we state our
> theorem, with the point in mind, that we wish to understand
> the continuity, of functions in the class $\mathcal{S}$.   ✠

We certainly use a comma to indicate a pause. But the comma indicates a *logical pause*, not a lack of air or lack of good sense. Read the last displayed sentence out loud, with suitable pauses where the commas occur. It sounds like someone huffing and puffing; the pauses have no reason to them. This sentence is not a representative example of the way that we speak, hence it is not indicative of the way that we should write. Much more attractive is

> Our hypotheses are now in place, and we next state our the-
> orem. The point is to understand the continuity properties
> of functions belonging to the class $\mathcal{S}$.

Mathematicians like the word "given". We tend to overuse and misuse it—especially in instances where the word can be discarded entirely. Consider the example "Given a metric space X, and a point $p \in X$, we see that ...". More direct is "If X is a metric space and $p \in X$, then ...". We are often tempted to transcribe spoken language and call that written language; such laziness should be defeated. Our misuse of "given" is an example of such sloth.

Whenever possible, use singular constructions rather than plural. Consider the sentence

> Domains with noncompact automorphism groups have orbit
> accumulation points in their boundaries.   ✠

First, such a construction is quite awkward: should it be "groups" or "group"? More importantly, do all the domains share the same automorphism group, or does each have its own? Does each domain have several orbit accumulation points, or just one? Clearer is the sentence

> A domain with noncompact automorphism group has an orbit accumulation point in its boundary.

When you are putting the final polish on a manuscript, look it over for general appearance. In mathematical writing, several consecutive pages of dense prose are not appealing, nor are several consecutive pages of tedious calculation. For ease of reading, the two types of mathematical writing should be interwoven. It requires only a small extra effort to produce a paper or book with comfortable stopping places on every page. The reader needs to take frequent breathers, to survey what he has read, to pause and look back. Make it easy for him to do so.

While you are thinking about the counterpoint between prose and formulas, think also about the use of displayed math versus in-text math (in TeX, the former is set off by double dollar signs $$ while the latter is set off with single dollar signs $). Long formulas are usually better displayed, for they are difficult to read when put in text. Of course *important* formulas should be displayed no matter what their length—and provided with numbers or labels if they will be mentioned later. Do not display every single formula, for that will make your paper a cumbersome read. Also do not put every formula in text, as that will make your writing tedious. A little thought will help you to strike a balance, and to use the two formats to good effect.

And now a coda on the role of English in mathematical writing. More and more, English is becoming the language of choice. Therefore those of us who are native speakers set the standard for those who are not. We should exercise a bit of care. I have a good friend, also an excellent mathematician, who is widely admired; his fans like to emulate him. He is fond of saying (informally) "What you need here is to cook up a function f such that ...". Mathematicians of foreign extraction, who have been hearing him make this statement for years, have now developed the habit of saying "Take a function f. Now cook it for a while ...". It is a bit like having your children emulate (poorly) all your bad habits. A word to the wise should suffice.

1.9 Pretension and Lack of Pretension

Avoid the use of big words when small ones will do. Do not say "peregrinate" for "walk", nor "omphaloskepsis" for "thought", nor "floccinoccinihilipilificate" for "trivialize" unless the longer word conveys some important nuance that the shorter word does not. The urge to so bloviate should be resisted. To indulge in hippopotomonstrosesquipedalian tergiversation is not to show your erudition; rather, it is to be superficial. Also remember that many of your readers will be foreign born, not native English speakers. Make some effort to write simple, straightforward English that they will easily apprehend. Save your high-flown rodomontade for ceremonial occasions.

Likewise—and I have said this elsewhere in the book—stick to simple sentence structures. Even the subjunctive mood can lead to confusion when it is used in mathematical writing. Let the mathematics speak for itself; do not try to dress it up with fancy language.

You can have some fun peppering your prose with *bon vivant* and *Gemütlichkeit* and *ad hominem* and *samizdat*, but the careless use of foreign words and phrases does not add anything to most writing. And it will confuse many readers. Use foreign phrases sparingly. If you do use them, typeset them in italics. (An exception should be made for foreign words like "etc." (short for *et cetera*), which have become standard parts of the English language and should be set in roman.) The books [Hig], [Por], [SG], [Swa] give more detailed treatments of this topic.

Good mathematics is difficult. Do not let your writing be a device for making it more so. Use simple, declarative sentences—short ones. Use short paragraphs, each with a simple point. To understand my meaning, put yourself in the position of the reader. You are slugging your way through a tough paper. You come to the proof of the main theorem. After killing yourself for a couple of hours, you finally come to the crux of the argument. And it is a single, dense paragraph that is two pages long. Such a daunting prospect is truly depressing. You do not want to abuse your readers in this fashion. Break up the ideas into palatable bites.

And now a note on flippancy. A friend of mine once wrote a truly elegant—and important—book that included the phrase "the reader should review enough functional analysis so that he does not barf [*sic*] at the sight of a Banach or Frechet space." At the reviewer's insistence, the phrase was toned down before publication. Another friend published a book with the phrase "we leave the details of this proof for the mentally infirmed." I would advise against this sort of sarcasm. This suggestion is not simply a nod to propriety. You want to be proud of your work. Remember that your thesis advisor and the authorities in the field are likely to look at it. Such puerile prose is not what you want them to see. Most likely, ten years hence, you will wish fervently that you had not included such phrases. Anyone who continues to grow intellectually will look on his work of ten years ago with some disdain. But there is no percentage in adding embarrassment to the mix.

Suit your tone, and your choice of words, to the subject at hand. It might be suitable to use the phrase "He had all the efficiency and dexterity of a ruptured snail" to describe a clumsy waiter; this is probably not appropriate language for describing the pope.

Finally, stay away from faddish prose. If you say "fraternally affiliated, ethically challenged young male" to mean "gang member" or "peregrinating, fashion-challenged, pulchritudinally advanced hostess" to mean "prostitute", then you may be politically correct today but you will be strictly out to lunch tomorrow. Today, many a writer or speaker wants to work the word "dis" (gang talk for "disrespect") or "flame" (yuppie talk for "disrespect") into his prose. This practice is a mistake, because in ten years the words will have no meaning.

By the same token, avoid old-fashioned modes of expression. In 1827 it was appropriate for a physician to diagnose a patient with "falling crud and palpitation of the pluck"; in 1930 it was fashionable for a woman to complain of "the fantods". Today these phrases are meaningless. It might exhibit devotion to Fermat to use "adæquibantur" instead of "=" (as did he), but such a practice would lead to boundless confusion today.

Some American writers think that it tony to pepper their writing with British English. They use "humour" for "humor", "lorry" for "truck", and "spanner" for "wrench". Such language is out of place, and can only lead to obfuscation. It would be just as foolish for an American cookbook to give recipes for spotted dick, bubble-and-squeak, and stodge. Nobody would know what the author was talking about.

For the same reasons I advise against using contractions, using abbreviations, or using slang—at least in formal writing. Even acronyms are dangerous (see Section 1.12); use them with caution. We write because we want our thoughts to last, and to be comprehensible both now and in years hence. Do not let language stand in the way of that goal.

1.10   We vs. I vs. One

When I was a child, I once asked a mathematician why mathematics was usually written in the first person plural: "We now prove this"; "Our next task is thus"; "We conclude our story as follows." The rejoinder that I received was "This is so that the reader will think that there are a lot of you."

More seriously, when you are writing up mathematics then you must make a choice. You can say "I will now prove Lemma 5" or "We will now prove Lemma 5" or "One may now turn one's attention to Lemma 5." Which is correct?

As with many choices in writing, this one involves a degree of subjectivity. I shall now tell you what I think about the matter. The first option is rarely chosen. Most people consider it pompous and inappropriate. The only instance where I find the first person singular to be a comfortable choice is the following: sometimes at the end of a paper one says "At this time I do not know how to prove Conjecture A." The choice is appropriate for this particular statement because in fact you are imparting to the reader some specific information about what you yourself know. It would be misleading, and a trifle affected, to say "At this time *one* does not know ...". Likewise for "At this time *we* do not know ...".

The custom in modern mathematics is to use the first person plural, or "we". It stresses the participatory nature of the enterprise, and encourages the reader to push on. Moreover, since "we" is what people are accustomed to hearing, it is less likely to jar their ears, or to distract them, than one of the other choices. The use of third person singular, or "one", often leaves the writer struggling with awkward sentence structures. If you endeavor to write in that mode, then you will likely find yourself soon breathing a sigh of relief as you abandon it. If you read with sensitivity, you also will likely learn that first person singular, or "I", is irritating; therefore you will not use it.

With a little craftsmanship, you can avoid entirely the use of the first person in your writing. Rather than say "We now turn to the proof of Lemma 4," instead say "Next is the proof of Lemma 4" or perhaps "The next task is the proof of Lemma 4." Rather than say "We see that the proof is complete," say "The proof is now complete" or "This completes the proof." The book [Dup, Ch. 2] has a sensible and compelling discussion of the question of "We" vs. "I" vs. "one".

Sound and sense will dictate which of the words "I", "we", or "one"—or perhaps none of these—you wish to use. I am offering "we" as the default. But the sense of what you are writing may dictate another choice.

1.11   Essential Rules of Grammar, Syntax, and Usage

I have intentionally put this discussion of the rules of grammar and syntax and usage at the end of Chapter 1. The reasons are several. First I want, in a gentle way, to de-emphasize them. I am not one of those who says "the battle against 'hopefully' is lost," "the battle for 'which' vs. 'that' is lost," "the battle for 'lay' vs. 'lie' is lost," and so forth. I find such statements facile, and they miss the point that careful writing requires some precision. The argument "You know what I mean; whether I use 'that' or 'which' is incidental" abrogates the fact that accurate writing, and accurate expression of your thoughts, requires accurate use of language. But you do not develop skill as a writer by concentrating on the rules of language; they are merely a set of tools that are used in the process.

The intent of this book is that you should learn to write logically and cogently; to say precisely what you mean, using just the right number of words; to eschew obfuscation. You want to develop an ear, so that clear writing becomes natural. Exact use of the language is a part of the process. But it is not the main point.

Most of the rules of English usage are succinct and logical. A particularly concise enunciation of the basic rules appears in [SW]. Since I cannot improve on that presentation, I certainly shall not repeat the rules of grammar here. It is a revelation for any adult writer to review the rules of usage and to see what eminent sense they make. Here I shall mention just a few sticky points that come up frequently in mathematical and other writing. I hope that you will find this section, and the next, to be a useful "quick-and-dirty" reference. With that goal in mind, I have presented the topics in alphabetical order. See also [Chi], [Dup], [Fow], [Fra], [Hig], and [MW] for a more thorough treatment of issues of grammar, syntax, and usage.

Bear in mind, as you read these precepts, that no rule of English grammar is etched in stone. There will certainly be times that a sentence or phrase formed according to the strictest rules will sound just awful. In such an instance, you must override the rules and use your good sense and taste. More will be said about this technique as the book develops.

Now for some rules:

- ***All, Any, Each, Every*** In mathematics we commonly formulate statements such as "Show that any continuous function f on the interval $[0, 1]$ has a point M in its domain such that $f(M) \geq f(x)$ for $x \in [0, 1]$." For cognoscenti it is clear that, when we say "any" here, we mean "all". But for others—for students, or for nonnative speakers—this slight abuse of language could cause confusion. For example, a student reading this sentence could (perfectly correctly) construe it to mean "Demonstrate that for *some* function f ...". Thus, if this sentence were part of an exercise, the student might answer

    > The function $f(x) = -(x - 1/2)^2$ is continuous on $[0, 1]$
    > and the point $M = 1/2$ satisfies the conclusion.

    The lesson is to avoid using "any" when "all" or "each" or "every" is intended.

    Conversely, even when you are writing for experts you can cause confusion by misusing quantifiers. Sentences like

    All continuous functions have a maximum.   ✠

are far too common in mathematical writing. Notice that the sentence suggests that all continuous functions share the *same* maximum. Of course what was intended was

    Every continuous function has a maximum.

or, more precisely,

    Each continuous function has a maximum.

(Once again we see the advantage, from the point of view of clarity, of the singular over the plural.) As you proofread your work, you must learn to take the part of the reader (who is not *a priori* sure of what is being said) in order to weed out misused quantifiers.

- **Brevity** Endeavor to formulate your thoughts briefly and succinctly. For example, you *could* say

    In point of fact, we devolved upon the decision to solicit opinions, form an enumeration, and produce a tally.

                                                        ✠

Such a sentence sounds mellifluous, sanguine, and high toned. But why not instead say

    We decided to take a vote.

The second sentence says in six words what the first said in 19; and it presents the message more clearly and forcefully. Strunk and White [SW] give a thorough and engaging treatment of the topic of brevity, and they speak particularly cogently of eliminating extra or extraneous words. Mathematics is difficult to read under the best of circumstances. Do not make the reader's job even more difficult by weighing down your prose with excess baggage.

- ***cf., e.g., i.e., n.b., q.v.,* and the like** These are abbreviations for specific Latin expressions: *confer*, *exempli gratia*, *id est*, *nota bene*, *quod vide*. They have particular meanings, and you should strive to use them accurately. In particular, "*cf.*" is often misused to mean "see". It actually means "compare". Sometimes "*e.g.*" and "*i.e.*" are interchanged in error; the first of these means "for example", and the second means (literally) "the favor of an example" or (more familiarly) "for the sake of example." It is difficult to use *n.b.* with grace. If you are unsure, then use the English equivalent of which you *are* sure.

  In fact it is difficult to make a compelling case for "*i.e.*" in favor of "that is", or for any of the other Latin substitutes in favor of their English equivalents. The punctuation and font selection questions connected with these Latin abbreviations are tricky (see [Hig] or [Fow] or [Chi] or [SK]). To repeat, use them with care.

- **Contractions** Do *not* use contractions in formal writing. Thus the words "don't", "can't", "shouldn't", "I'm", "you're", etc., are taboo. Of course you should never write "ain't". You also should avoid abbreviations. Particularly avoid using informal abbreviations like "cuz" for "because", "tho" for "though", and so forth. You will probably never be tempted to work "bar-b-q" into your next paper on para-differential operators; but you might be tempted to use "rite inverse". Please resist.

  Occasionally you will find it suitable to use contractions in various kinds of *informal* writing. It can be a way of drawing in your audience, or of warming yourself up to your subject. For example, in the book [Kr2] I intentionally used an occasional contraction in an effort to create a friendly air about the book. By contrast, the present book is a book about writing, and I wish to set a more formal example—so there are no contractions.

- ***Denote*** Use the word "denote" carefully. It has a special purpose in mathematics (and in logical positivist philosophy and modal logic) and we should take care to preserve it as a useful tool. Suppose that a certain mathematical symbol A stands for, or represents, the item or set of ideas B (ideally, you should be able to excise any occurrence of A and replace it with B and preserve exactly the intended meaning). Under these circumstances, and *only under these circumstances*, do we say that "A denotes B." For example,

> Let X denote the set of all semisimple homonoids with stable quonset hut.

There is a shade, but an important shade, of difference between the statements

> **(1)** Let f be a continuous function.

and

> **(2)** Let f denote a continuous function.

The intended meaning here is "let f be *any old* continuous function." Thus the first statement is both customary and correct. The second is neither customary *nor* correct. For we use "denote" when we want to say that a certain specific item stands for some other specific item. That is not what we are trying to say here.

Lack of familiarity with English, or lack of familiarity with the precise meaning of "denote", sometimes leads to dreadful abuses of the word. A common one is "Denote X the set of all left-handed polyglots." I leave it to you to decide whether failing English or failing intellect might be the correct provenance of such a sentence; the lesson for you is not to use "denote" in such a fashion.

The word "connote", rarely used in mathematical writing, can be (but should not be) confused with "denote". The dictionary teaches us that "A connotes B" means that A *suggests* B, but not in a logically direct fashion. For example,

> To a young man, "love" connotes flowers, beautiful music, and happiness.

is an appropriate use of the word "connote".

- *If* vs. *Whether* The words "if" and "whether" have different meanings, and are suitable for different contexts. Follow the example of master editor George Piranian:

  > Go to the window and see *whether* it is raining; *if* it is raining, then let Fido inside.

- *Infer* and *Imply* The words "infer" and "imply" are often confused in everyday usage. It should not be difficult for a mathematician to keep these straight. A set of assumptions can *imply* a conclusion. But one *infers* the conclusion from the assumptions. It is that simple.

- *Its* and *It's* Use "it's" only to denote the contraction for "it is". Otherwise use "its". For example "Give the class its exam" and "A place for everything and everything in its place." Compare with "It's a great day for singing the blues."

  More generally, the apostrophe is never used to denote the possessive of a pronoun: what is correct is "its", "hers", "his", "theirs".

- *Lay* and *Lie* "Lay" is a transitive verb and "lie" is intransitive. This means that "lay" is an action that you perform on some object, while "lie" is not. For instance, "Lay down your weary head," "Now I will lay down the law," and "I shall lay responsibility for this transgression at your feet"; compare with "I am tired and I shall lie down" or "Let sleeping dogs lie." Note, however, that the past tense of "lie" is "lay". Therefore you may say "Yesterday I was so tired that I laid down my books and then I lay down."

- *Less* and *Fewer* How many times have you been in the grocery store and gravitated toward the line that is labeled *Ten Items or Less?* Of course what is intended here is *Ten Items or Fewer*, and I have a special place in my heart for those few grocery stores that get it right. The word "fewer" is for comparing two numbers while "less" is for comparing quantity. Mathematics deviates a bit from these rules, because we certainly say "3 is less than 5" (of course the meaning here is that "the number 3 represents a quantity that is less than the quantity represented by the number 5"). Avoid saying "3 is smaller than 5," because "smaller" is a word about *size*: perhaps the digit 3 is smaller than the digit 5. It also *could* be correct to say "5 is smaller than 3" if comparison of digit size is what is intended.

- **Lists Separated with Commas (the Serial Comma)** When you are presenting a list, separated with commas, then you should put a comma after every item in the list except the last. For example, say "the good, the bad, and the ugly" rather than "the good, the bad and the ugly." A moment's thought reveals that the former conveys the intended meaning; the latter may not, for the reader could infer that "bad" and "ugly" are somehow linked.

- **Numbers** Some sources will tell you that (whole) numbers less than 101 should be written out in words; larger numbers should be expressed in numerals (other sources will put the cutoff at twenty or some other arbitrary juncture). A discursive discussion appears in [SG]. Such considerations are, for a mathematician, next to ludicrous. The main thing, and this advice applies to spelling and to many other *choices*, is to select a standard and to be consistent.

- ***Obviously, Clearly, Trivially*** These words have become part of standard mathematical jargon. This is too bad. In the best of circumstances, when you use these phrases you are endeavoring to push the reader around. In the worst of circumstances you are throwing up a smoke screen for something that you yourself have not thought through. It would be embarrassing to count the number of major published mathematical errors that have been prefaced with "Obviously" or "Clearly" (no doubt the supreme deity's way of reminding us that "pride goeth before the fall").

The use of these words is one of the ways that we have of kidding ourselves.

As you proofread your manuscript relentlessly, and endeavor to weed out superfluous words, pay particular attention to the use, abuse, and overuse of these trite words. They add nothing to what you are saying, and are frequently a cover-up.

- **Overused Words** Certain words in the English language are grossly overused. Among these are "very" and "most" and "nice" and "interesting". It is certainly very pleasant and most insightful to express great appreciation for a very nice and supremely interesting theorem; but I encourage you not to do so—at least not with these banal words. If such language represents how you wish to express yourself, then perhaps you have nothing to say. Instead think carefully about what you really mean to say, and then say it.

  In fact the language is littered with overused words that come into and out of fashion. The words "awesome", "totally", "dude", and "righteous" are current examples. The phrase "today I'm not 100%," foisted upon us by some semiliterate sports announcer, is currently the bane of our collective existence. Each field of mathematics has its own set of stock phrases and tiresome clichés. Endeavor not to propagate them.

  A good general principle is to put every word in every sentence under the microscope: What does it add to the sentence? Will the sentence lose its meaning if the word is omitted? Can the thought be expressed with fewer words? Strunk and White [SW] have a splendid discussion of the concept of weighing each word.

- **Plural Forms of Foreign Nouns** We all grind our teeth when we hear our freshmen say "And this point is the *maxima* of the function." To no avail we explain that "maxima" is plural, and "maximum" is singular. Yet we make a similar error when we do not differentiate "data" (plural) from "datum" (singular) and "criteria" (plural) from "criterion" (singular). As usual, exercise special care when dealing with foreign words.

- **The Possessive** When you express the possessive of a singular noun, always use 's. Thus you should say "Pythagoras's society", "the dog's day", "Stokes's theorem", "Bliss's book", "baby's bliss", and "van der Corput's lemma". The terminal "s" is omitted when you are denoting the possessive of a plural noun: "the boys' trunk", "the dogs' food", "the students' confusion".

  "Collective nouns" are treated in a special manner. For instance, we write "the people's choice" and "the children's folly": even though the nouns are plural, we denote the possessive *with* a terminal "s".

- **Precision and Custom** At times, the goal of precision in writing flies in the face of custom. Antoni Zygmund once observed that the World Series of American baseball might more properly be called the "World Sequence". I am inclined to agree (in no small part out of fealty to my mathematical grandfather), but I must be over-ruled by custom: if you use the phrase 'World Sequence' then nobody will know what you are talking about. Bear this thought in mind when you are tempted to invent new terminology or new notation (see also the remarks in Section 2.4 on terminology and notation).

- **Subject and Verb, Agreement of** Make sure that subject and verb match in your sentences. A mismatch not only grates on the sensitive ear, but can seriously distort meaning. Consider the example "The set of all morphisms are compact." This syntax is incorrect. The *subject* (that is, the person or thing performing the action) in this sentence is *set*. We should conjugate the verb "to be" so that it agrees with this subject. As a result, the grammatically correct statement is "The set of all morphisms is compact." (Note, in passing, that the original form of the sentence might have misled the reader into thinking that the writer was—rather clumsily—discussing a collection of compact morphisms.)

Of course the test is easy: omit the prepositional phrase "of all morphisms" and analyze the root sentence. Clearly "The set is compact" is correct while "The set are compact" is not. You will find the device of focusing on the root statement, or breaking into pieces (see our analysis of Subject and Object below), to be a valuable tool in analyzing many grammatical questions.

As a parting exercise, consider that "the sequence $\{z_n\}$ *converges* to p" while "the numbers z_n *converge* to p". Think carefully about why both statements are correct.

- *This* and *That* We often hear, especially in conversation, phrases like "Because of this, we decided that." If we exercise the full force of logic then we must ask " 'Because' of *what*?" and " 'we decided' *what*?" And this niggling query raises an entire body of common errors that I would like to point out. This corpus is not composed so much of errors in English usage, but rather errors in logic and precision. Consider the following examples:

Shakespeare was an important writer. This tells us a lot about English literature.   ✠

A triangle is a three-sided polygon. This means that . . .   ✠

The day was bright and beautiful. Because of this, Mary smiled.   ✠

In each of these sample sentences, my objection is " 'this' what?" (Notice that I did *not* say "In each of these, my objection is . . .". I was careful to say *precisely* what I meant.) The following passages convey the same spirit as the preceding three, but they actually *say* something:

Shakespeare was an important writer. The forms of his plays and poems as well as his use of language have had a strong influence on English literature.

A triangle is a three-sided polygon. The trio of sides satisfies the important *triangle inequality*.

The day was bright and beautiful. Observing the weather caused Mary to smile.

Here is a delightful example that was contributed by G. B. Folland: "Saddam Hussein was determined to resist attempts to force Iraqi troops out of Kuwait, although George Bush made it clear that he did not want to be seen as a wimp. This caused the Gulf War." If you were to ask someone to which clause "this" refers, then the answer you received would probably depend on that person's politics.

The message here is fundamental: as a default, do not use "this" or "that" or "these" or "those" without a clear point of reference. When the occurrence of "this" or "that" is fairly close to the referent, then the intended meaning is often clear from context. When instead the distance is greater (as in Folland's example), then confusion can result.

Repetition is a good thing, so repeat your nouns rather than refer to them with a potentially vague pronoun or other word. There *will* be cases where the casual use of "this" or "that" is both natural and appropriate, but such instances will be exceptions.

Copy editor Rosalie Stemer says that a hallmark of good writing is that it answers more questions than it raises. Applying this philosophy will lead naturally to many of the points raised in this book, including the present one.

- **Where** One of the most common types of run-on sentence in mathematics is a statement with a dangling concluding phrase such as "where A is defined to be ...". An example is

    Every convex polynomial function is of even degree, where we define a function to be convex if ...   ✠

We see this abuse so often that we are rather accustomed to it. This is also an easy crutch for the writer: he did not bother to plant the definition before this statement, so he just tacked the definition onto the end.

That is sloppy writing and there is no excuse for it: before you use a term, define it. You need not use a formal, displayed definition. But you must put matters in logical order. The example I have given is quite trivial; but in serious mathematical writing it is taxing on the reader to have to pick up definitions on the fly.

- ***Who** and **Whom**;* **Subject and Object** Be conscious of the difference between "who" and "whom". The word "whom" is an object; used properly, it denotes a person that is *acted on*. An example of the common misuse of the word "whom" is

> The pastor, whom expected a large donation, smiled warmly.   ✠

Here the issue is what is the correct subject to put in front of the verb "expected". The word "whom" cannot act as a subject. The correct word is "who": "The pastor, who expected a large donation, smiled warmly." In the same vein, it is correct to say "To whom am I speaking?" and "Is he the man to whom the Nobel Prize was awarded?"

Also do not confuse "I" and "me". The latter is an object, the former not. For example, "The teacher was addressing Bobby and I" is plainly wrong, since here "I" is used incorrectly as the object of the verb "addressing". President Clinton's famous misstatement "Give Al Gore and I a chance to bring America back" is a dreadful error; nobody would say "Give I a chance . . . ". That sort of sentence analysis is the method you should use to detect the error. The sentence

> Him and me proved the isotopy isomorphism theorem in 1967.   ✠

is an abomination. Unfortunately, even smart people make mistakes like this. Anyone can see that "Him proved the isotopy isomorphism ..." and "Me proved the isotopy isomorphism ..." are incorrect. But, somehow, the ganglia are more prone to misfire when we put the two sentences together. Conclusion: test the correctness of a sentence with compound subject (or any compound element) by breaking it into its component sentences.

1.12   More Rules of Grammar, Syntax, and Usage

Here I include additional rules of grammar and syntax that are dear to my heart. They come up frequently in general writing, less so in specifically mathematical writing. They should prove useful in your expository work, and sometimes in your research work as well.

- **Adjectives vs. Adverbs** An adjective is designed to describe, or to modify, a noun. An adverb is designed to describe, or to modify, a verb. Correct is to say

  This is a good book.

and

  This is an expensive car.

and

  The quick, brown fox jumped over the stupid, lazy dog.

because "good", "expensive", "quick", "brown", "stupid", and "lazy" are adjectives. They modify the nouns "book", "car", "fox" (twice), and "dog" (twice), respectively. You may also say

  He shouts loudly.

and

He sings beautifully.

and

He strove sporadically to master his homework thoroughly.

because "loudly", "beautifully", "sporadically", and "thoroughly" are adverbs. They modify the verbs "shouts", "sings", "strove", and "master". Learn to distinguish between adjectives and adverbs, and learn to use both correctly.

- **Alternate vs. Alternative** The words "alternate" and "alternative" (used as adjectives) have different meanings, though they are often, and erroneously, used interchangeably. The word "alternate" (most commonly used in the form "alternately") refers to some pair of events that occur repeatedly in successive turns; the word "alternative" refers to a choice between two mutually exclusive possibilities. For example:

  Pierre alternately dated Mimi and Fifi. He had considered monogamy, but had instead chosen the alternative lifestyle of a concupiscent lothario.

- **The Verb *To Be*** "The verb 'to be' can never take an object." Probably you have been hearing this statement all your life. What does it mean?

  When you formulate the sentence

  I hit the ball.

  then "I" is the subject (of the verb "hit") and "ball" is the object (of the verb "hit"). But when you formulate the sentence

  I am the walrus.

  then "I" is the subject (of the verb "to be", conjugated as "am") but "walrus" is the *predicate nominative* (also sometimes called the *predicate noun* or *subjective complement*). The word "walrus" is *not* the object of an action. It has a different grammatical role in this sample sentence.

By the same token, it is incorrect to answer the query (over the telephone) "Is this Napoleon Bonaparte?" with the answer "This is me." The word "me" is supposed to be used as the object of an action. The verb "to be", however, does not take an object. Thus the correct rejoinder is "This is I" or "This is he."

To make a long story short, your writings should not include the statement "The person who proved Fermat's Last Theorem is me." Grammatically correct is "The person who proved Fermat's Last Theorem is I" or "It is I who proved Fermat's Last Theorem" or "I am the one who proved Fermat's Last Theorem." You should not, however, pen any of these statements unless you are Andrew Wiles.

- ***Compare* and *Contrast*** The words "compare" and "contrast" have different meanings. One compares two items in order to bring out their similarities; one contrasts two items in order to emphasize their differences. For instance, we can compare groups and semigroups because they are both associative. We can contrast them because one contains all inverse elements and the other does not.

- ***Different from* and *Different than*** The phrase "different from" is correct, while "different than" is not. Examples are "His view of grammar is different from mine" and "His syntax is different from what I expected." The classical rationale here is that the word "different" demands a preposition and a noun. Modern treatments (see [Fra, p. 266]) suggest that this classical dictum is too restrictive and that "different than" (without the noun) is more comfortable. You will have to decide which usage you prefer, but do be consistent.

- ***Due to, Because of,* and *Through*** Mathematicians commonly use the phrase "due to", and we often use it incorrectly. We sometimes say "due to the fact that" when instead "because" will serve nicely. The phrase "due to" tempts us to wordiness that is best resisted.

A good rule of thumb (thanks to G. Piranian) is to use "due to" only to introduce an *adjectival clause*—never an adverbial clause. In fact the grammatical issues at play here—including the use of "through"—are rather complex, and not suitable for discussion in this book. See [MW] for a detailed treatment.

- *Farther* **and** *Further* It is common to interchange the words "farther" and "further", but there is a loss of precision when you do so. The word "farther" denotes distance, while "further" suggests time or quantity. For example, one might say "I wish to study *further* the question of whether Lou Gehrig could throw the baseball *farther* than Ty Cobb."

- **Good Taste and Good Sense**   Suit your prose to the occasion. The writer of a Harlequin romance novel might write

    Clutched in the gnarled digits of the syphilitic Zoroastrian homunculus was a dazzling Fabergé egg.

  while Raymond Chandler would have written something more like

    The dwarf held a gewgaw.

  In mathematics, simpler is usually better. Flamboyant writing is out of place.

- *Hopefully* **and** *I hope* With due homage to Edwin R. Newman [New], I note that it is incorrect to use "hopefully" (at the beginning of a sentence) when you mean to say "It is hoped that" or (more sloppily) "I hope". The word "hopefully" is an adverb. It is intended to modify a verb. For example,

    She wanted so badly to marry him, and she looked at him hopefully while she waited for a proposal.

Note that the word "hopefully" modifies "looked". It is incorrect to say

Hopefully the weather will be better today.   ✠

when what you mean to say is

I hope that the weather is better today.

By the same token, do not say "This situation looks hopeful." People can be hopeful, objects or things never.

Monty Python tells us that "Mitzi was out in the garden, hopefully kissing frogs." If you are comfortable with the common misuse of "hopefully" then you will probably misunderstand this sentence.

The reference [KnLR, p. 57] offers a detailed analysis of the history of the word "hopefully", and another, more liberal, point of view about its use. See also [MW].

- **Infinitives, Splitting of** As a general rule, do not split infinitives. For example, do not say "He was determined to immensely enjoy his food, so he smothered it in ketchup." The correct version (though one may argue with the sentiment) is "He was determined to enjoy his food immensely, so he smothered it in ketchup." Here the infinitive is "to enjoy" and the two words should not be split up. Curiously, the reason for this rule is an atavism: some of the languages that contributed to the formation of modern English, such as Latin and French, combine these two words into one. Our rule not to split the infinitive carries on that tradition.

There are a number of opinions on this matter. The "modern" point of view is that it is acceptable to split an infinitive when it sounds right; otherwise it is not. For example, sometimes a mathematical sentence will resist the suggested rule. G. B. Folland supplies the example "Hence we are forced to severely restrict the allowable range of values of the variable x." Strictly speaking, the word "severely" splits the infinitive "to restrict". But where else could you put "severely" while maintaining the precise meaning of the sentence?

On a more personal level, the sentence

> To really love someone requires a lot of effort.

evinces a particular sentiment while

> To love someone really requires a lot of effort.

conveys a dissimilar meaning.

Arguably, it would be better to formulate the sentence differently (how about "To love a person with passion and intensity requires a lot of effort"?). But if one were wedded to the "really" construction then one would have a problem. Use your ear, and use sound and sense, to convey your message clearly and forcefully.

- ***In terms of*** Sentences of the form

> Who is he, in terms of surname? ✠

and

> How is he doing, in terms of his math classes? ✠

are simply dreadful. Usually the phrase "in terms of" is gratuitous, and can be omitted entirely. Consider instead

> What is his surname?

and

> How is he doing in his math classes?

- **Need Only; Suffices to**   In written mathematics, we often find it convenient to say "We need only show that ..." or "It suffices to show that ...". These are lovely turns of phrase. Strive not to overuse them, or to misuse them. Too often we see instead "We only need to show that ..." or "Suffices it to show that ...". With these misuses, the message still comes across—but in a more halting and less compelling manner.

- **Parallel Structure**   The principle of parallel structure is that proximate clauses which have similar content and purpose are (often) more effective if they have similar form. The use of parallel structure is an advanced writing skill: good writing can be made better, more forceful, and more memorable with the use of parallel structure. Consider the dictum

    Candy is dandy but liquor is quicker.

  Whether you approve of the sentiment or not, the thought is memorably expressed—using a quintessential example of parallel structure. As an exercise, try expressing the thought with more desultory prose, and see for yourself what is lost in the process.

    The first inspirational quotations (from Sir Francis Bacon) in Chapters 3 and 5 provide less frivolous examples of parallel structure.

- **Participial Phrases**   Participial phrases are a frequent cause of discomfort. For example,

    Shining like the sun, the man gazed happily upon the heap of gold coins.   ✠

  The participial phrase "shining like the sun" modifies "man", whereas it was clearly intended to modify "the heap of gold coins." Better would be

    The man gazed happily upon the heap of gold coins, which shone like the sun.

Harold Boas contributes the following useful maxim: "When dangling, don't use participles."

- **Prepositions, Ending a Sentence with** As a general rule, do not end a sentence with a preposition. Do not say "Where do we stop playing at?" Instead say "At what point do we stop playing?" Better still is "When do we stop playing?" Do not say "What book are you speaking of?" Instead say "Of which book do you speak?" or "Which book is that?"

Often, when you are tempted to end a sentence with a preposition, what is in fact occurring is that the errant preposition is a spare word—not needed at all. The preceding examples, and the suggested alternatives, illustrate the point.

An old joke has a yokel trying to find his way across the Harvard campus. A Brahmin student corrects him sternly for posing the question "Excuse me. Where's the library at?" After the Harvardian explains at length that one does not end a sentence with a preposition, the yokel tries again: "Excuse me. Where's the library at—*jerk*?" This is perhaps a bizarre example of sound working with sense.

- **Quotations** We do not often include quotations in mathematics papers. If you decide to include a quotation, then be aware of the following technicality. Logically, it makes sense to write a sentence of the following sort:

> As Methuselah used to say, "When the going gets tough, the tough get going".

What is logical here is that the quotation itself is a proper subset of the entire sentence; therefore it stands to reason that the terminal double quotation mark should be *before* the period that terminates the sentence. Unfortunately, logic fails us here. Admittedly typesetters are still debating this point, but the current custom is to put the closing double quotation mark *after* the period. Open any novel and see for yourself. Thus the sentence *should* be written

> As Methuselah used to say, "When the going gets tough, the tough get going."

The fact is that the complete rule is even a bit more complicated than has already been indicated. By the rules of *American* usage, commas and periods should be placed inside quotation marks, and colons and semicolons outside quotation marks (see [SG, p. 222] and [Dup, p. 192]). Placing exclamation points and question marks inside or outside of quotation marks depends on context. British usage is even more ambiguous. This is all a bit like the infield fly rule in baseball. But do be consistent, and be prepared to arm-wrestle with your publisher or with your copy editor if you have strong opinions in the matter.

We sometimes find it awkward to follow the rules of the last paragraph in every instance. For example, consider the sentence

> Let us now discuss the usage of the word "pulchritude".

Here I violate the stated dictum and put the period *outside* the closing quotation mark. I do so because the quotation marks surround a single word (or short phrase, without a verb). This custom is more natural, and more pleasing to the eye, than the standard rule cited for our friend Methuselah's statement.

If your quotation is n paragraphs in length, then there is an opening double quotation mark on every paragraph. There is no closing double quote on paragraphs 1 through $(n - 1)$; but there certainly *is* a closing double quote on paragraph n.

- **Redundancy** Logical redundancy, used with discretion, can be a powerful teaching device. By contrast, avoid (local) verbal redundancy. The phrases "old adage", "funeral obsequies", "refer back", "advance planning", "strangled to death", "invited guest", "body of the late", and "past history" display an ignorant and superfluous use of adjectives. Avoid constructions of this sort.

- *Shall* and *Will* In common speech, the words "shall" and "will" are often used interchangeably, or according to what appeals to the speaker. In formal writing there is a simple distinction: when expressing belief regarding a future action or state, "shall" is used for the first person ("I" or "we") and "will" is used for the second person ("you") or third person ("he", "she", "it", or "they"). When the first person is expressing determination, then it is appropriate to use "will". These rules, taken from [SW], are illustrated whimsically in that source by

  Bather in Distress: "I shall drown and no one will save me."

but

  Suicide: "I will drown and no one shall save me."

- *That* and *Which* The word "that" is used to denote *restriction* while the word "which" denotes *amplification*. For example, "I am speaking of the vase that sits on the table" and "The book that is by Gibbons is in the study." Compare with "The vase, which is red, sits on the table" and "The book, which is by Gibbons, is fascinating."

  Here is a mathematical example: "A holomorphic function that vanishes on S must be identically zero." Compare with "A holomorphic function which vanishes on S must be identically zero." Which is correct? Think about the logic. What we are saying is that a holomorphic function f such that $f(z) = 0$ for $z \in S$ must be identically zero. (For the mathematics, note that in one complex variable a set S with an interior accumulation point will suffice for the truth of the statement.) Phrased in this way, the statement is restrictive: a holomorphic function with a certain additional property must be zero. Thus the correct choice is "that" rather than "which".

Modern grammarians approve of the use of "which" for "that" in suitable contexts. Consult a grammar book, such as [SG], for the details.

I have already noted that it is sometimes useful to let "sound and sense" overrule the strict code of grammar. In particular, there are times when "which" sounds more weighty, or more formal, than "that". Thus some writers will make the technically incorrect choice, just to achieve a certain effect.

As already noted, the rules of grammar and syntax are not absolute. English usage is constantly evolving. While some current aspects of usage are fads and nothing more, others become common and are finally adopted by the best writers and speakers. Those tend to stay with us. But there is a more subtle point. Sometimes a sentence formed according to the strict rules of usage *sounds awkward*. A classic example (usually attributed to Winston Churchill) is

That is the sort of behavior up with which I will not put.

Notice that the speaker is going into verbal contortions to avoid ending the sentence with a preposition. The result is a sentence that is so ludicrous that it defeats the main purpose of a sentence—to *communicate*. Better is to say

That is the sort of behavior that I will not put up with.

While technically incorrect—because the preposition is at the end of the sentence—this statement nevertheless will not grate on the ears of the listener, will convey the sentiment clearly, and will get the job done. Of course it would be even better to say

I will not tolerate that sort of behavior.

This sentence conveys exactly the same meaning as the first two. But it has the advantage that it is direct and forceful. In most contexts, the last sentence would be preferable to the first two. This is again a matter of sound working with sense. And here is a point that I will make several times in this book: often it is a good idea *not* to wrestle with a sentence that is not working; instead, reformulate it. That is what we did with the last example.

As an exercise, find a better way to express the following sentence (which ends with five prepositions, and which I learned from Paul Halmos by way of [KnLR]):

> What did you want to bring that book I didn't want to be read to out of up for?

Do not use acronyms, abbreviations, or jargon unless you are dead certain that your audience knows these shortcuts. Speaking of an ICBM, the NAFTA treaty, ARVN, and MIRV is fine for those well read in the current events of the past twenty-five years—and who have an excellent memory to boot. But most of us need to be reminded of the meanings of these acronyms. The best custom is to define the acronym parenthetically the first time it is used in a piece of writing. For example,

> The SALT (Strategic Arms Limitation Talks) were progressing poorly, so we broke for lunch. A few hours later, we resumed our efforts with SALT.

These days I am on many AMS (American Mathematical Society) committees, and am somewhat horrified at the extent to which I have become inured to acronyms. How many of these do you know: CPUB, COPROF, JSTOR, LRPC, ECBT, COPE? I am conversant with them all, and none has done me a bit of good. In practice, you may not even safely assume that your reader knows what the AMS is—what if he is Turkish?

I was once at a meeting to discuss the writing of a new grant proposal—to apply for renewal of funds from a generous source which, we hoped, would be inclined to give again. One of the PI's ("PI" denotes "Principal Investigator") said, in all seriousness, "I think that we are going to need more blue sky in this proposal if we want to generate more bottom line." Of course his meaning was "We must endeavor to paint an enlarged picture of long-term goals and anticipated achievements if we want to increase the size of this grant." The first mode of expression might be appropriate among venture capitalists, who are inured to such language. It is probably inappropriate among academics.

Chapter 2
Topics Specific to the
Writing of Mathematics

What I really want, doctor, is this. On the day when the manuscript reaches the publisher, I want him to stand up—after he's read it through, of course—and say to his staff: "Gentlemen, hats off!"

Albert Camus
The Plague

You don't write because you want to say something; you write because you've got something to say.

F. Scott Fitzgerald

So I'm, like, "We need to get some food." And he's, like, "I don' wanna go th' store. How 'bout some 'za?" And I'm, like, "Well, we gotta eat, dude. I could get like totally into a pizza." And he's, like, "No biggie." And I'm, like, "This guy is grody to the max. Gag me with a spoon."

A Valley Girl

We have read your manuscript with boundless delight.
If we were to publish your paper,
it would be impossible for us to publish any work of lower standard.
And as it is unthinkable that in the next thousand years
we shall see its equal, we are, to our regret,
compelled to return your divine composition and to beg
you a thousand times to overlook our short sight and timidity.

Memo from a Chinese Economics Journal
From *Rotten Rejections* (1990)

Having imagination, it takes you an hour to write a paragraph that, if you were unimaginative, would take you only a minute. Or you might not write the paragraph at all.

Franklin P. Adams
Half a Loaf (1927)

2.1   How to Organize a Paper

To begin, a mathematics paper has certain technical components. It requires a title, and that title should convey some information to the reader. If it does not, then the reader is likely to move on to more stimulating reading matter, without looking any further at your work. Thus a title like *On a theorem of Hoofnagel* says almost nothing. The title *On differentiation of the integral* is only slightly better; but at least now the reader knows that the paper is about analysis, and he has a rough idea what sort of analysis. The title *Quadratic convergence of Lax/Wendroff schemes with optimal estimates on the error term* is ideal. In a nutshell, this title tells the reader exactly what the paper is about and, further, what point it makes.

Of course an equally important component of your paper is the identification of the author or authors. At the beginning of your career, pick a name for yourself and stick to it. And I do not mean a name like "Stud" or "Juicymouth". I might have called myself Steven George Krantz or S. G. Krantz or S. Georgie Krantz or any number of other variants. I chose Steven G. Krantz, just as it appears on the title page of this book. When an abstracting, indexing, or reviewing service endeavors to include your works, you want it to be a zero-one game: it should retrieve all your works or none of them. You do not want any to be left out, and you should leave no doubt as to your identity.

Here is a quick run-down of other technical components that belong in most papers: **(1)** affiliations of authors, **(2)** postal addresses and *e*-mail addresses of authors, **(3)** date, **(4)** abstract, **(5)** key words, **(6)** AMS subject classification numbers, **(7)** thanks to granting agencies and others. Items **(1)**, **(2)**, and **(3)** require no discussion; **(4)** is discussed in Section 2.5. Let us say a few words about **(5)** – **(7)**.

The key words are provided so that *Math. Reviews* and other archiving services can place your paper correctly into a database. Endeavor to choose words that reveal what your paper is about; that is, you want words that will definitely lead a potential reader to your paper. Thus "new", "interesting", and "optimal" are not good choices for key words. Instead, "pseudoconvex", "Cauchy problem", and "exotic cohomologies" *are* good choices.

Similar comments apply to the AMS subject classification numbers. The American Mathematical Society has divided all of mathematics into 94 primary classification areas (rather like *phyla* in the classification of animals) and these in turn into subareas. Assigning the correct classification numbers to your paper is a reliable way to put your paper before the proper audience. It also will help to ensure that your paper is classified correctly. The American Mathematical Society publishes an elegant little book, which can be found in most mathematics libraries, that lays out the AMS subject classification scheme. The scheme also appears in certain issues of *Math. Reviews*. The key words and classification numbers usually appear in footnotes on the first page of your paper; some journals instead specify that title, abstract, key words, and classification numbers appear on a separate "Title Page".

Finally item (**7**): often it will be appropriate to thank other mathematicians for helpful conversations or specific hints. Sometimes you will thank someone for reading an earlier draft of the paper, or for catching errors. If you want to do things strictly by the book, you should ask a person before you thank him in public (because, for example, most people would not want to be thanked heartily in a paper that turned out to be hopelessly incorrect). However, as a matter of fact, most people do not engage in this formality; and most of those who are thanked do not object. Occasionally you may wish to thank the referee for helpful comments or suggestions (best is to do this *after* the paper has been refereed—not in advance, or in anticipation of a friendly referee); sometimes you will need the editor's help in handling this particular "thank you" correctly. Sometimes one thanks one's spouse for forbearance, or one's typist for a splendid job with the manuscript. Sometimes one thanks one's department for time off to complete the work, or for the opportunity to teach an advanced seminar in which the work was developed. The one particular form of thanks that you are honor bound to include is thanks to any agency—government, university, or private—that has provided you financial (or other) support. In some cases, this thanks is mandatory; in all cases it is an eminently appropriate courtesy.

Now let us turn to the contents of the paper. A mathematical paper is not a love letter to yourself (in content it might be, but in form it definitely should not be). You are writing about a topic on which you have become expert. You have made an advance, and you want to share it with the mathematical community. This should be your point of view when writing up your results.

The simplest way to write a paper is to introduce some notation, state your theorem, and begin the proof (for simplicity I am supposing that this is a "one-theorem paper"). Such a procedure probably involves the least effort on your part, it gets the theorem recorded for posterity, and it might even get the theorem published. But this methodology is the least effective if you genuinely want your work to be read and understood, and if you want the ideas disseminated to the broadest possible audience.

In point of fact a good mathematics paper is *not* necessarily written in strict logical order. The reason lies in theories of learning due to Piaget and others. The point is simply this: While it can be useful—when recording mathematics for the record—to develop ideas à la Bourbaki/Hilbert in strict logical order, *this is not the way that we learn.* It is not the way that a typical human being—even a mathematician—apprehends ideas. This is the case even if the reader is an expert in the subject, just like yourself.

Reading a mathematics paper is work, and a typical reader approaches the task with caution. Most people will not read more than a couple of math papers per month—I mean really read them, verifying all the details. However, those same people will *look* at several dozen papers each month. We all receive a great many preprints in the mail and over the Internet. We must make choices about which ones to *read*.

Having established this premise, let us think about what sort of paper will encourage the potential reader to plunge in, and what sort will not. If the first couple of pages of the paper consist of technical definitions and technical statements of theorems, then I would wager that most potential readers will be discouraged. Imagine instead a paper written as follows. The first paragraph or two summarizes the main results of the paper, in nontechnical language. The next several paragraphs provide the history of the problem, describe earlier results, and state exactly what progress the current paper represents. This introduction concludes, perhaps, with acknowledgements and an outline of the organization of the paper (either in Table of Contents form or paragraph form).

A reader faced with the latter organizational form has many advantages. This person knows **(i)** what the paper is about, **(ii)** why the result of the paper is new, **(iii)** what is the context into which the paper fits, and **(iv)** whether he wants to read on.

One person whom you must keep in sharp focus as you craft your paper is the referee. You cannot, indeed you must not, assume that the referee will compensate for your shortcomings. If *you* do not explain what the paper is about, why you wrote it, why your theorems are new, why this paper makes an interesting contribution, why its techniques are original—then nobody else is going to do it for you. And the referee, who really does not want to do the full job of reading the *entire paper*, will (if the introductory portion of your paper is not up to snuff) conclude quickly that this is just another piece of second rate drivel and will reject it.

Back to the chase: Imagine that, having concluded the introductory section of the paper, you (the writer) turn to the necessary technical definitions and a formal statement of results. This central material would be the substance of Section 2 of the paper (assuming that the introductory material, discussed in the last paragraphs, was Section 1). Now the reader—the expert who has desired to slog this far—knows precisely what he is getting himself into. Turning to Section 3, you (the writer) can now dive into all the pornographic details of the proof. Right? Wrong.

Reading a difficult mathematical proof in strict logical order is an onerous task. If the first five pages of Section 3 consist of a great many technical lemmas, with nary an indication of where things are going, of what is important, and of what is not, then many readers will be discouraged. Let me now describe a better way.

It is more work for the writer, but definitely a great favor to the reader, to organize the paper as follows. Section 3 should consist of the "big steps" of the proof. Here you should formulate the technical lemmas (provided that the reader can understand them at this point) and then you should describe how they fit together to yield the theorem. You should push the nasty details of the proofs of the lemmas to the end of the paper.

It should be clear by now why the proposed organizational scheme makes sense. First, the reader can decide at each of these signposts how far he wants to get into the paper. Each new epsilon of effort on his part will yield additional and predictable benefit. And the hardest and most technical parts are left to the end for the real die-hard types. This writing style is of course beneficial for the reader; it will also aid the writer. It disciplines the writer, forces him to evaluate and predigest what he has to say, and will tend to reduce errors.

The principles of writing a math paper that have been described here do not apply to every paper that is, has been, or ever will be written. They probably do not literally apply to the Feit/Thompson paper on the classification of finite, simple groups (an entire issue of the *Pacific Journal*) or to Andrew Wiles's proof of Fermat's Last Theorem (an entire issue of the *Annals*). They certainly apply to a twenty-page, "one-theorem" paper. And the general principles described above probably apply in some form to virtually any mathematics paper.

And now a word about redundancy. In general, redundancy is a good thing. One fault that all mathematicians have is this: we think that when we have said something once clearly then that is the end of it; nothing further need be said. This observation explains why mathematicians so often lose arguments. You must repeat. Help the reader by recalling definitions—especially if the definition was given 50 pages ago. If you need to use the definition *now*, and if you have not used it for quite a while, then give the reader some help. Give a quick recap or at least a cross-reference; do likewise for a theorem or a lemma that you need to recall. Think of how much you would appreciate this assist if you were the reader.

2.2   How to State a Theorem

There are some mathematical subjects—geometric measure theory is one of them—in which the custom is for the statement of a theorem to occupy one or more pages, and for the enumerated hypotheses to number 25 or more. This practice is too bad. It makes the subject seem impenetrable to all but the most devoted experts. People who present their theorems in the fashion just described are wont to claim that their subject prevents any other formulation of the theorems, that this is just the nature of the beast. I would like to take this opportunity humbly to disagree. It may require extra effort on the part of the writer, but I claim that you never need to state a theorem in this tedious manner.

You should strive to hold the statement of a theorem to fewer than ten lines, and preferably to five lines. (Some books on writing assert that a theorem should consist of only a single sentence!) How can you do this if there are twenty-five hypotheses? First of all, the assertion that there are twenty-five hypotheses is only a manifestation of what is going on in the writer's mind. Mathematical facts are immutable and stand free from any particular human mind, but the way that we describe them, verify them, and understand them is quite personal. In particular, the way that a theorem is presented is subject to considerable flexibility and massaging. Let us consider a quick and rather artificial example:

Theorem: Let f be a function satisfying the following hypotheses:

1. The function f has domain the real number line;

2. The function f is positive;

3. The function f is uniformly continuous;

4. The function f is monotone;

5. The function f is convex;

6. The function f is differentiable except possibly on a set of the first category;

7. The function f has range that is dense in the positive real numbers;

8. The function f has no repeated values;

9. The function f is a weak solution of the differential equation
$$Lf = 0$$
(where the operator L has been defined earlier in the paper);

10. The function f^2 is a subsolution of $Lf = 0$.

Then f operates, in the sense of the functional calculus, on all bounded linear operators on a separable, real Hilbert space H.

This sample "theorem" has only ten hypotheses, and these assumptions are not all that difficult to absorb; but it serves to illustrate our point. Here is a more efficient, and more user-friendly, manner in which to state the theorem.

Suppose that, prior to the statement of the theorem, we formally define a function to be *regular* if it is defined on the real line, uniformly continuous, convex, monotone, and positive. Further, we define a function to be *amenable* if it has range dense in the positive reals and has no repeated values. Finally, let us say that a function f is *smooth* if it is differentiable except possibly on a set of the first category, it is a weak solution of L and, in addition, f^2 is a subsolution of L. Each of these should be stated as a formal definition, prior to the formulation of the theorem. Moreover, we should state that, until further notice, H will designate a separable, real Hilbert space and $\mathcal{L}(H)$ the bounded linear operators on H. With this groundwork in place, we can now state the theorem as follows.

Theorem: If f is a regular, amenable, smooth function, then it operates on $\mathcal{L}(H)$ in the sense of the functional calculus.

Notice that, by planning ahead and introducing the terms "regular", "amenable", and "smooth", we have grouped together cognate ideas. We are not just engaging in sleight of hand; in fact we are providing organization and context. We are also helping the reader by keeping the statement of the theorem short and sweet. The reader will come away from reading the theorem remembering that **(i)** there is a hypothesis about f involving continuity, convexity, and so forth, **(ii)** there is a hypothesis about the value distribution of f, and **(iii)** there is a hypothesis about the way that L acts on f. The conclusion is that f operates on $\mathcal{L}(H)$. You, the writer, have done some of the work for the reader, and given him something to take away. The reader can always refer to the text for details as they are needed. But if the theorem is recorded in the first form rather than the second then, most likely, the reader will not quite know what he has read, nor when and where he can use it.

Also note that we managed to state the theorem in one sentence, and in just two lines.

2.3 How to Prove a Theorem

What I mean here, of course, is "how to *write the proof of a theorem*." You are not doing your job—unless the proof is short and fairly simple— to begin at the beginning and charge through to the end. A proof of more than a few pages should be broken into lemmas and corollaries and organized in such a fashion that the reader can always tell where he has been and where he is going.

A useful device in writing up a proof is the "Claim". This tool is often used in the following manner. You have set up the basic pieces of your proof; that is, you have defined the sets and functions and other objects that you need. You are poised to strike. Then you write "We claim that the following is true." State the claim. Then you say "Assuming this claim for the moment, we complete the proof."

Used correctly, this technique is a terrific psychological device. It allows you to say to the reader "This is the crux of the proof, but its verification involves some nasty details. Trust me on this for the moment, and let me show you how the crux leads to a happy ending." The reader, having arrived at the end of the proof (modulo the claim), will feel that progress has been made and he will be in a suitable mood either to study the details of the claim or to skip them and come back to them later.

Another useful device—nearly logically equivalent to the "claim"— is to enunciate a technical lemma right at the point where you need to use it (sometimes a good idea because to enunciate it well in advance would make almost no sense to the reader), but then to say "Proof Deferred." If you indulge in this trick, be sure that your paper is well organized and that the different parts of the paper are well labeled. Do not leave your poor reader with a head full of dangling claims and unproved lemmas to sort out. A good rule of thumb ([Gil, p. 8]) is to be sure that your reader always knows the *status* of every statement that you make.

In Section 2.1 I have advocated that a paper should be organized so that the technical stuff is at the back and the explanatory stuff at the front. The paper should proceed, by gradations, from the latter to the former. The proof of a theorem should proceed in roughly the same way. You, as the author and creator of the theorem, have the whole thing jammed into your head; it has no beginning and no end—it just resides there. Part of the writing process is to transfer this organic mass from your head to someone else's. Thus, as you write, try to provide signposts so that the reader always knows where he has been and where he is going. This writing goal is best achieved by pushing the technicalities to the end.

Many books, and some papers, are written as follows: the author rattles on for several pages—chatting about this and that—and abruptly says

> Note that we have proved the following theorem:
>
> **Theorem [The Riemann Mapping Theorem]:** Let Ω be a simply connected, proper subset of the complex plane . . . .   ✠

Good heavens! What a disservice to the reader. The Riemann mapping theorem is a milestone in mathematical thought, perhaps even in human thought. Each of the steps in its proof—the extremal problem, the normal families argument, etc.—is a subject in itself. The writer must lay these milestones out for the reader and must pay due homage to each. The offhand "Note by the way that we have proved the Riemann mapping theorem" is a real travesty, and ignores the author's duty to *explain.* Rise above the idea that it suffices for the writer to somehow record the thoughts on the page; if you, the author, have not crafted them and worked them and, indeed, handed them to the reader, then you have not done your job.

And here is a small note about proofs by contradiction. Some mathematicians begin a proof by contradiction with

> Not. Then there is a continuous function f . . .   ✠

Others begin with

> Deny. Then there is a continuous function f ...   ✠

This is all rather cute; the first of these is perhaps a tribute to John Belushi and the *Saturday Night Live* gang. But both examples (and these are *not* made up—people actually write this way) hinder the task of *communicating*. A preferred method for beginning a proof by contradiction is

> Seeking a contradiction, suppose that f is a continuous, real-valued function on a compact set K that does not assume a maximum. Then ...

2.4 How to State a Definition

Definitions are part of the bedrock of mathematical writing. Mathematics is almost unique among the sciences—not to mention other disciplines—in insisting on strictly rigorous definitions of terminology and concepts. Thus we must state our definitions as succinctly and comprehensibly as possible. Definitions should not hang the reader up, but should instead provide a helping hand as well as encouragement for the reader to push on.

As much as possible, state definitions briefly and cogently. Use short, simple sentences rather than long ones. To avoid excessively complex and introspective definitions, endeavor to *build* ideas in steps. For instance, suppose that you are writing an advanced calculus book. At some point you define what a function is. Later you say what a continuous function is. Later you state what the intermediate value property for continuous functions is. Still later, you use the latter property to establish the existence of $\sqrt{2}$. You do not, all at once, attempt to spit out all these ideas in a single sentence or a single paragraph. In fact you build stepping stones leading to the key idea, so that the reader is given a chance to internalize idea n before going on to idea $(n+1)$.

Just how many definitions should you supply? If you are writing a paper on von Neumann algebras (algebras of bounded operators on

Hilbert space), then you certainly need not say what a Hilbert space is, nor what a bounded linear operator is. Every graduate student who has passed through the qualifying exams is familiar with these ideas, and you may take these for granted. (That is why we have qualifying exams.) Define $\mathcal{L}(H)$ (see Section 2.2) only if you think that readers likely will misinterpret this (rather standard) notation. Of course you would have to define "regular", "amenable", and "smooth" (the terminology that we introduced in Section 2.2). Those terms are not standard, and have been given other specialized meanings elsewhere.

What I am describing here is another of many subjective matters that pertain to writing. If your paper supplies too few, or poorly written, definitions then both the referee and the readers will lose their patience. If your paper supplies too many definitions, then you also will irritate your audience. For standard terminology, you could give a well-known reference like Dunford and Schwartz [DS] or Griffiths and Harris [GH] or Birkhoff and MacLane [BM] or Kuratowski [Kur]. This habit is preferable to taking up valuable journal space with a rehash of well-known ideas. Less kind is to refer to a semi-obscure journal article for terminology. If that is the best reference for definitions, then you should probably repeat them.

There is some terminology that you simply cannot take the space to repeat or define, even though it is rather advanced. For example, you cannot rehash—for the convenience of your readers—the standard theory of elliptic partial differential equations, nor the basics of K theory, nor the guts of the Atiyah-Singer Index Theorem. (In writing a book you in fact *can* indulge in such a review; I treat book writing elsewhere.) Try to refer the reader to a good source for the important ideas on which you are building.

I have advocated (Section 2.2) the tasteful use of terminology to clump ideas together, thus making them more palatable for the reader. However, try to avoid introducing any more new terminology than is necessary. If your paper contains a plethora of unfamiliar language, then it may cause your reader to suspect that you actually have nothing to say. And if there is a standard bit of notation or terminology for what you are saying, then by all means use it. I once saw a paper in a standard mathematics journal of good repute that defined the space Q^{17}_{reg} to be the set of all bounded holomorphic functions on the unit disc in the complex plane. Of course the well-known notation $H^\infty(D)$ describes this space of functions, and it is virtually mandatory to use *that* notation. The proposed alternative notation is just crazy, *unless* the author is introducing a whole new scale of function spaces in which H^∞ arises in a natural way. If that is the case, then the author should certainly mention this relationship explicitly. (For example, all the standard function spaces—L^p, Lipschitz, Sobolev, Hardy, Besov, Nikol'skii, etc.—are special cases of the Triebel-Lizorkin spaces $\dot{F}^{\alpha,q}_p$. Thus, in certain contexts, it would be appropriate to refer to the Lebesgue spaces, or the Sobolev spaces, using the Triebel-Lizorkin notation.)

Good notation is extremely important, sometimes as important as a theorem. As an example, the notation of differential forms is a small miracle. Large parts of geometric analysis would be completely obscure without it. Of course you cannot perform at the level of Elie Cartan every time you dream up a piece of notation, but you can consider following these precepts: **(1)** Do not create new notation if there already exists well-known notation that is suitable for the job at hand; **(2)** If you must introduce new notation, then think about it carefully; **(3)** Strive for simplicity and clarity at all times.

Fiddle with several different notations before you make a final decision. Consult the standard references in the field to see whether they give you any ideas. If possible, try your new notation out on a colleague, or on one of your graduate students.

Technically speaking, a definition should almost always be formulated in "if and only if" form. For example

> A function f on an open interval I is said to be *continuous* at $c \in I$ if and only if for every $\epsilon > 0$ there is a $\delta > 0$ such that ...

In practice, we generally replace the phrase "if and only if" in this definition with "if". We do so partly out of laziness, and partly because the "if" phraseology is less cumbersome than "if and only if". The price that we pay for this convention is that we must teach our students to read definitions; the fact is that we *do not* write what we mean.

Although nobody will punish you for writing "if and only if" in your definitions, and some will appreciate it, it is usually best to follow mathematical custom and simply to write "if". A useful, and modern, compromise is to use Paul Halmos's invention "iff" (see Section 1.8). The word 'iff" captures the brevity of "if" but carries the precision of "if and only if".

2.5   How to Write an Abstract

Many journals now require that, when you submit a paper, you include an abstract of the paper. The abstract, usually not more than ten lines, is supposed to convey on a quick reading what the paper is about. According to the strictest standards, the abstract should be self-contained, should not make any bibliographic references, and should contain a minimum of notation and jargon.

A rough rule of thumb is that any reader who looks at your paper will read the abstract, only 20% of those will read the introduction, and perhaps one fourth of that 20% will dip into the body of the paper. This being the case, your abstract is obviously of preeminent importance. Many indexing and reviewing services will rely on your abstract. So it had better give a clear picture of what is in the paper.

As usual, endeavor to employ simple, short, declarative sentences in your abstract. Eschew nasty details. Do not say, with a plethora of ϵ's and δ's, exactly what interior elliptic estimate you are proving; instead state that you are proving a new interior elliptic estimate in the Nikol'skii space topology and that it improves upon classical results of Nirenberg. State that it has applications to certain free boundary problems. The interested reader can then move on to the introduction, where further details are provided.

If your abstract is too long or too short, then the editor of the journal will likely make you rewrite it. The "Instructions to Authors" section in the journal should give you an idea of what is required for an abstract in *that* journal. Study several abstracts in the journal to which you plan to submit to get an idea of what is suitable.

2.6 How to Write a Bibliography

The bibliography, or list of references, is one of the most important components of a mathematical work. This assertion is true for research articles, for books, and for expository articles as well. The bibliography tells the reader where you are coming from and where you are going, it keeps you honest, and it provides critical assistance for those readers not already familiar with the subject.

Real sticklers—mavens of good scholarly form—will tell you that a bibliography should *only* be assembled from primary sources. The book [Hig, pp. 87–8] has several examples of bibliographic inaccuracies in the literature that have been propagated for dozens of years because reference $(n + 1)$ was always copied from reference n. The book [Hig] also advises you never to retrieve information about a paper either from the cover of the journal issue or from the Table of Contents since information is frequently misrepresented in both places. Even if it is not, you could easily get the first or last page of an article wrong. Accuracy and scholarship are best served when you gaze upon the actual paper; and you will also be able to say truthfully that you have "looked" at the paper.

Sticklers also will tell you that each reference should include an **MR** (or *Math. Reviews*) number. Such an addition is often quite convenient for the reader, and a lot of extra work for the writer (though with the advent of MathSciNet, the web service available from the American Mathematical Society, the task has become much easier).

You should only list references in your bibliography that you also cite in the text. We are frequently tempted to include extra references either for sentimental reasons or because we think that these references might be handy for the reader. The former motivation is spurious, and the latter misguided. If you give the reader no advice on the value of a reference, then you are offering nothing by listing it.

There are many possible formats for bibliographic entries. If you use AMS-TEX, then your bibliographic entries are formatted for you automatically in the approved AMS style. (Similar comments apply to LATEX's treatment of bibliographic entries.) But if you do not then you must make some choices. At the beginning of my career I picked a favorite journal and adopted its bibliographic style. I chose a for-

mat that is commonly used, and it has served me well. Here are two
bibliographic references formatted in that style:

[Bat]   Gill Bates, *How I Made My First Billion*, 2nd Ed.,
         Acquisitive Press, New York, 1986.

[Beh]   Viscount Hugh Behave, Some theories on the gentle
         art of belching in public, *The Journal of Eminently
         Forgettable Theories* 42(1976), 35-53.

The first of these is a book, and the second a paper in a journal. Notice
that the information provided for a book is different from that provided
for an article. For a book, the author, title, edition number (if this is
not the first edition), publisher, city of publication, and date of publication are usually considered complete bibliographical data. These are
shown, in order, in the example given. For a paper, the author, title, journal, volume number of the journal, year, and pages are usually
considered complete bibliographical data. Of course the protocol for a
preprint, for a conference proceedings, for an unpublished manuscript,
for a translated paper, and for a Ph.D. thesis are all a bit different. I
shall not go into all the details here. (See [SG, pp. 407-410] or [Hig] or
[VanL].)

   Not everyone likes the use of acronyms for citing elements of the
bibliography. Some people prefer to number the elements of the bibliography from [1] to [n]. The method of enumeration has the disadvantage that, if you add or delete a reference late in the game, it throws off
all your numbering. However using good software can circumvent that
problem (see below). Both the numbering scheme and the acronym
scheme have the disadvantage that even a one-character typographical error can make it virtually impossible for the reader to tell which
reference was intended.

   One excellent scheme for bibliographic references, and one that is
virtually essential when the bibliography is long, is illustrated in the
following example. It lists three works by John Q. Public, just as they
might appear in a bibliography.

John Q. Public

[1987] *Why I Never Vote*, Ignoramus Press, Brooklyn.
[1992a] *The Less I Know, the Better*, Rosicrucian Press, Poughkeepsie.
[1992b] On Doctoring Polls, *The Smart Pollster* 31, 59-71.

If you use this system (known as the *Harvard system*), then when you refer to a bibliographical item in the text you say "By J. Q. Public [1992b], we know that ...".

Note that in mathematics we do not usually put bibliographical references in footnotes (however it *is* customary in certain statistical work). This habit came about in part because typesetters objected to the expense and trouble of typesetting copious footnotes. With the advent of TeX, that particular objection is moot. However, the rule persists. In fact, if you were to submit to most mathematics journals a paper with all the references in footnotes, then you would most likely be asked to reformat it.

If you are writing your paper in LaTeX, then you have the option of using LaTeX's bibliographic utilities. One of LaTeX's tools allows you to assign a nickname to each of your bibliographical references. Then, in the text, you can cite any reference by its nickname. When you compile your *.TEX file, each nickname citation is replaced by the appropriate preassigned acronym or number; the full bibliographic citation occurs at the end of the document as usual (see Section 6.5 for more on TeX and LaTeX).

Slightly more sophisticated is LaTeX's bibliographic database system. With this device, you never write another bibliography as such. You simply have an ever-growing database of bibliographic references. Whenever a new reference comes to hand, you add it directly to the database. Each reference has a preassigned acronym and a preassigned nickname. Then, when you are writing a new document, you make a reference by referring to the appropriate nickname in the database (if you cannot remember all the nicknames—perhaps your database has thousands of items in it!—you can just pull the database into a window with your text editor and check it). When you compile the document, a beautiful bibliography is created for you, with the requisite information pulled in from the database.

If the last systems do not appeal to you, then you also can keep the TEX files of all your papers in a single directory. Most of us tend to use many of the same references repeatedly. Thus, when you are writing a new paper and need a reference, you can open a window with your text editor, pull in an earlier paper that has the reference, and cut and paste the reference into your new document.

Incidentally, LATEX also allows you to assign nicknames to your equations and theorems. You can refer to them, during the writing process, by nickname. Then, when the paper compiles, the correct line numbers and theorem numbers are inserted for you automatically.

I know mathematicians—excellent ones—whose bibliographies look like this:

1. Knuth, 1992.

2. Lister, 1991.

3. Machedon, 1988.

This is it! No titles, no journal names, no volume numbers, no page references. This scheme in effect takes the LATEX device to the limit: you just supply the nicknames but none of the details.

The practice of listing abbreviations in lieu of correct bibliographic references is irresponsible. In truth, such sloppiness should have been caught by the editor, who should have demanded that the author rectify the matter. As indicated at the beginning of this section, the bibliography is part of your paper trail. You hold the responsibility for providing complete bibliographic information. It should be complete in the sense that you have cited everyone who merits citation, but it also should be complete in the sense that all the information is there. The bibliographic sample just provided might mean something to a few experts for a few years. In fifty years it would not mean anything to anyone.

And speaking of "meaning nothing to anyone", do *not* give in-text bibliographic references that have the form "see Dunford and Schwartz" (for those not in the know, [DS] is a three volume work totaling more than 2500 pages). The only correct and thorough way to give a reference is to cite the specific theorem or the specific page. Sometimes, to conserve space and to prevent repetition, we say "by a variant of theorem thus and such" or "by a variant of the argument in this paper" (the subject of analysis, in particular, seems to be littered with references of this nature). If you find such references necessary in your own work, be as specific as you can so that the reader may follow your path.

Modern technology enables a marvelous writing environment—at least in principle. If, for example, I am a Windows95® user, then I can have my text editor going in one window (this is where I actually do my writing), a thesaurus and dictionary on CD-ROM in another, the library's on-line catalog in another, and MathSciNet on line in a fourth. Passing from one environment to the next requires only a keystroke or a mouse click or two. Clearly such an environment makes tedious trips to the library a thing of the past, and makes assembling a bibliography relatively quick and easy.

Now let us treat styles for citations. In this section, I have spoken of bibliographic references with the assumption that they will occur on the fly, right in the text. For example:

> By a theorem of Steenrod [Ste], we know that every instance of generalized nonsense is a generalization of specific nonsense.

The good feature of this methodology is that it tells you right away what the source is. The bad feature is that it clutters up the text a bit. In most mathematics *papers*, the on-the-fly style is used. You make a reference either by acronym, or by number, or by author surname, but the reference occurs at the moment of impact.

In [Ste], Steenrod fulminates against this bibliographic style for the writing of a book. His preference is to have a paragraph or more at the end of each chapter detailing the genesis, development, and sources for the theorems in that chapter. Many books in the Princeton book series *Annals of Mathematics Studies* handle bibliographic references in this fashion. These little end-of-chapter essays can be quite informative and, if well written, can give the reader a sense of the historical flow of thought that in-context references (as indicated above) do not. I would say that the down side of this end-of-chapter approach is the following. It serves the big shots well. If you are annotating a chapter on singular integrals, then you will certainly not overlook Calderón,

Zygmund, Stein, and the other major figures. But you might overlook the smaller contributors. The advantage of the in-text, on the fly reference method is that it systematically holds you accountable: you state a theorem, and you give the reference; you recall an idea, and you give the reference. You are much less likely to give someone short shrift if you adhere to this more pedestrian methodology. Of course the final choice is up to you.

2.7 What to Do with the Paper Once It Is Written

Ours is a profession where, by and large, we are left on our own to figure out how to function. Nobody shows us how to teach, nobody tells us how to write a paper, and nobody tells us how to get published. This section addresses the last issue.

So imagine that you have written a paper that you think is good. How do you know it is good? Being a mathematician is a bit like being a manic depressive: you spend your life alternating between giddy elation and black despair. You will have difficulty being objective about your own work: before a problem is solved, it seems to be mightily important; after it is solved, the whole matter seems trivial and you wonder how you could have spent so much time on it. How do you cut through this imbroglio?

If you are smart, you have told some colleagues about your results. Perhaps you have given some seminars about it. You have sent preprints around (either by *e*-mail or by snail mail) to colleagues. If you have kept your ears open, you have some sense of how receptive the world is to your ideas. Are your listeners surprised, impressed, confused, bored? Sometimes they will suggest changes. Consider all criticisms and suggestions carefully, and make appropriate changes to your paper. Now you must decide where to submit it.

Before you make that momentous decision, let me back-pedal a minute and address the question of how to decide when you have something that is worth writing up. This is a confusing issue, and one that every mathematician must learn to face.

We all know that the keys to success in this profession of ours include intelligence, perseverance, drive, and hard work (not necessarily in that order). Some may deny it, but there is also an art to the business. Let me explain. Ideally, the working mathematician sets a problem for himself: solve the (restricted) Burnside problem, or calculate the dual of the Hardy space H^1, or prove the corona theorem in several variables. We all know that there are extraordinary mathematicians who can actually do just this: E. Zelmanov did the first and C. Fefferman did the second. Nobody has done the third, although many of us have tried. In practice, this point-and-shoot technique is rarely the way that mathematics is successfully practiced.

A somewhat more modest way to get one's feet wet is this: become completely immersed in a subject, and then formulate a program. Determine to assume hypotheses A, B, C and endeavor to prove conclusion X. Sadly, this *modus operandi* is also only occasionally successful.

In fact what happens in practice is that we try a great many things. Some succeed and some do not. Along the way, hypotheses are constantly being altered and substituted and strengthened; conclusions are redirected or transmogrified or reversed. The theorem that you end up proving is rarely the theorem that you set out to prove. This is a perfectly reasonable way to proceed. Columbus sought a new passage to India and instead found America. Jonas Salk discovered the polio vaccine by accident. Milnor discovered multiple differentiable structures on the 7-sphere because calculations on another problem were not working out as planned.

One of the chief differences between a successful mathematician and an also-ran is that the former can take his partial results and his tries—and yes, even his failures[1]—and turn them into an attractive tapestry of theorems and corollaries and partial results and conjectures; the latter instead takes two years of hard work and dumps it in the trash.

As you read these words, do not suppose that I am advocating any degree of chicanery, or self-promotion, or hype. I am instead encouraging you to have the confidence and fortitude to make something of your work. Part of doing mathematics successfully is to get in there

---

[1] A twentieth century Hungarian philosopher once said that a mathematician is nothing but a collection of statements that he cannot prove.

and calculate and reason and think and ponder. But another part is to evaluate and organize and deduce. What I am describing is a bit easier to imagine for a laboratory scientist. He performs a huge experiment that may take a year or two and may cost a few million dollars. No matter how things turn out, he must make a show of it. He must report to his granting agency and write papers about how his laboratory has been spending its time and effort. The message here is that a mathematician must do something similar, but his wherewithal is somewhat more tenuous; indeed it is all in his head. Part of training yourself to survive in this profession is coming to terms with the reality that I have described.

I cannot conclude this digression without also noting that another key to success is actually making some progress. It just will not do to tell yourself (and the world) that for the next twenty years you will work on the Riemann hypothesis, *unless you can arrange to have something to show along the way.* You do not get tenure, or a promotion, or an invitation to the International Congress by advertising that you are working on a great problem and telling people that they should contact you a generation later to see how things worked out. I have a friend who has a twenty-five step program for proving the Riemann hypothesis: "Count to twenty-four and then prove the Riemann hypothesis." There is wisdom in this little joke. The successful mathematician knows how to manage his research program so that it will proceed incrementally, so that he can report progress along the way—including writing up papers and giving talks and showing the world just what he is up to. By the same token, the good mathematician knows how to determine when he is *not* making progress, when his program is *not* paying off, when it is time to move on to something else.

Now let us return to more pedestrian matters. Let us suppose that you have organized some of your material and turned it into a paper. You believe that this is a worthy piece of work. You want to get it published. The next move is yours.

Keep in mind that the one hard and fast rule in this business is that you can submit a paper to just one journal at a time. *Never consider deviating from this policy.* In the words of Clint Eastwood, "Don't even think it." If you do send the same paper to two different journals simultaneously, then that paper is liable to be sent to the same referee by both journals; thus you will be caught red handed! Agonizing though it may be, you must wait for a decision from journal n before you submit to journal $(n+1)$. As a result, there is considerable motivation to exercise wisdom when choosing a journal.

There is a distinguished mathematician, now retired, who in his heyday wrote about a dozen papers per year. He submitted them all to *The Annals of Mathematics*. Several of his papers were accepted by the *Annals*. Others were either rejected or else the author was asked to perform various revisions. Now, writing twelve papers per annum as he did, this mathematician had no time for revisions. So, in cases two and three, he sent his papers to a well-known journal that was reputed to have minimal standards (what the famous computer scientist Dijkstra would call a "write-only" journal). Thus this esteemed man has a publication list, emblazoned in *Math. Reviews* for all to see, consisting of several citations in the *Annals* alternating with citations in this other "catch-all" journal.

Another famous mathematician was in the habit of bringing his latest preprint to the departmental secretary, together with a list of journals to which it might be submitted. Her job was to cycle through the journals on the list, one by one, and to inform Herr Doktor Professor when his paper was finally accepted. In this way the good Doktor was spared the grief of dealing with surly referees and uncooperative editors.

The preceding two strategies are amusing but probably unwise for most mathematicians. The working mathematician should have a sense of which are the very best journals, which are at the next level, and which are of average quality. How can one gauge which journals are which? They all look rather elegant, and all profess to have high standards. They all have distinguished people on their editorial boards. What is the trick?

Begin by considering where cognate results have appeared. The *Journal of Algebra* will probably not consider papers on singular integrals. The *Journal of Symbolic Logic* probably does not publish papers on Gelfand-Fuks cohomology. Certain journals have become the default forum for work on operator theory or several complex variables or potential theory. Consider those if your work fits. You will naturally consider which editors will understand what your paper is about and will know how to select a referee. You need not actually *know* the editor, but it is comforting to know where the editor is coming from.

If you submit your work to a journal of the highest rank, then you might pay in several ways: **1)** the refereeing process may take an extra long time, **2)** the journal may have a huge backlog, **3)** the paper may be rejected for almost any reason. Thus the entire process of getting your work published could drag on for two years or more. If you are fighting the tenure clock, this could be a problem. In some ways it is better to err on the low side. Usually mathematical work is judged on its own merits. Nobody will downgrade your work, or you, if your theorems are not published in the optimal journal. But do not publish in an obscure journal that nobody ever reads.

Part of the secret to success in this profession is to talk to people. Doing so, you will quickly learn that *Acta Mathematica*, the *Annals*, *Inventiones*, and the *Journal* of the American Mathematical Society are four of the pre-eminent mathematics journals. This choice of four reflects my prejudices as an analyst. Others might name *The Journal of Differential Geometry* or *The Journal of Algebra* or the *The Journal of Symbolic Logic* as being at the top. Opinions will vary. Perhaps *Duke*, the *Transactions* of the AMS, and several others are at the next level. And on it goes. There are prestigious journals and there are excellent journals. Many journals fit into both categories, and many fit into neither.

You can form your own opinion of journals by seeing what papers they publish and by which authors; you can look at how many truly eminent people (and from which universities) are on their editorial boards, and you can learn something just by submitting your papers to various journals and seeing what happens.

Since the latter strategy is costly—in terms of time, and perhaps your bruised feelings as well—you should develop a sense of what is a typical *Annals* paper, what is a typical *Transactions* paper, and what is a typical *Rocky Mountain Journal* paper. If you are in doubt, ask someone with more experience. If someone whom you respect and trust has read your preprint, then he would be an ideal person to ask for suggestions as to where to submit.

Also of interest in considering journals is the backlog of the publication, the turnaround time from submission to acceptance (or rejection), and similar data. Fortunately, the *Notices* of the AMS publishes a detailed review—containing just this sort of information—of all the major journals at least once per year. In the end, you have the responsibility to pick a suitable journal for your work; and the choice is not a trivial matter, since a year may pass while you are waiting for an acceptance or rejection.

Of course you will learn from experience. You also will have to decide for yourself whether to shoot high and take your chances, or to shoot low and optimize your likelihood of a quick acceptance. If your tenure case is a few years down the road, then this choice should not be taken lightly. Deans tend to know which are the good journals and which are not. (In fact I know of several universities where the dean has circulated a ranked list of mathematics journals. The implication is that "If you want to get promoted then you had better publish in these journals but not in those journals.") They are not impressed by a young assistant professor whose work is all submitted to "gimme" journals. They are also not impressed by a dossier with most papers "submitted" but not yet accepted.

Most journals have a section called "Instructions to Authors" or "Instructions for Submission". Before you submit paper X to journal A, you should read those instructions. They will tell you how many copies are needed, whether the title and abstract and other data should be on a separate page, whether the journal requires key words and AMS subject classification numbers, what languages the journal will accept (English, French, and German are the most common—though there *are* mathematics journals that will take papers in Latin or Esperanto or Japanese), any formatting requirements, length restrictions, where to send the paper (to the Editorial Office, or to an Associate Editor of your choosing, or perhaps another option), whether the journal prefers submissions in TeX, whether the journal has a TeX style file that you should use, whether the journal accepts electronic submissions, and so forth. You will annoy the editors, and cause unnecessary delays and confusion, if you do not follow these readily available instructions.

Before you submit your paper to a journal, please proofread it, spell-check it, and then do so again. Flip through each copy that you are submitting to ensure that each page is there, is right side up, and that the printing appears on the front of each sheet of paper. Few things are more irritating to an editor, or to a referee, than to receive a sloppily prepared manuscript.

Always send in all the requisite copies of your paper, together with a rather explicit cover letter, *in a single envelope*. The cover letter should say something simple like

> I wish to submit the paper "On the Cohomology of Proofs," by C. Gauss, D. Hilbert, and H. Poincaré, to *The Utopian Journal of Hemi-Semi-Demi Theorems*. Enclosed please find three copies. Thank you for your attention.

Be sure to put the title and complete list of authors of the paper into the cover letter. Should the cover letter become separated from the manuscript, its contents would be useless if it lacked this information.

The journal will assume that the "communicating author" is the person signing the cover letter *at the address indicated on the letterhead*. All further correspondence will be conducted with that person at that address. You should supply your *e*-mail address and telephone number and fax number in the cover letter. If you want someone else to be the communicating author, or if you want to specify a different address, then do so explicitly in the cover letter.

Some authors, endeavoring to be thrifty, send the cover letter by first-class mail and the manuscripts separately by library rate. This is a mistake, since the journal office does not know, in the absence of a cover letter, what it is receiving. Worse, suppose that you send the manuscript, without a cover letter, directly to one of the Associate Editors. He might open the envelope and think that he has just received a preprint to read. It will be thrown into his preprint pile, and soon forgotten. And you will wonder what happened to your paper!

Some authors think that the cover letter is an opportunity to make a pitch for the paper. Such an author will fill the cover letter with fulsome praise of what is in the preprint, why it improves on the existing literature, and who might be a suitable referee. Most editors will not find such remarks helpful, and many will find them annoying. By naming potential referees, you may in fact be ruling them out in the mind of the editor (since he may think that they are your pals). Best is to keep the cover letter simple and dispassionate.

Here is what you can expect after you have sent your paper (usually in duplicate or triplicate—although there are journals that request as many as five copies) to a journal. Within four to six weeks, the journal will notify you that it has received the paper; for legal reasons, this is usually done in writing (with a postcard, for instance), although some journals will just use *e*-mail. The journal will often assign a manuscript number to your paper, and will advise you to use this number in all future correspondence. I run a journal, and I can tell you that this number is valuable. The journal office can easily misfile a paper with multiple authors; also, since the paper is passed from Managing Editor to Associate Editor to one or more referees, the paper can be misplaced. It helps significantly when authors and editors use the manuscript number. Such a number might be "94-63", indicating that this is the 63rd manuscript received in 1994. The postcard will conclude by saying something like "Don't call us; we'll call you." In other words, you may have to wait a while for the referee's report; so sit tight.

Expect to wait four to six months for a report. After that wait, you are well within your rights to send a polite note (either by *e*-mail or by snail mail) to the editor to whom you submitted the paper; simply state that you submitted the paper on thus and such a date, received an acknowledgement on another date, and you are wondering if there has been any progress in the matter. Most editors appreciate a gentle reminder, and will in turn nudge the appropriate Associate Editor or referee.

Eventually you will receive a referee's report. It may be a paragraph or it may be five pages or more. It may say "This paper is terrific. Publish it as quickly as you can." Or it may say "This paper is dreadful. Stay as far away from it as you can." Most often it will say something in between these extremes.

If the paper is rejected, then you will have to ply your wares elsewhere. A rejection does not necessarily mean that the paper is bad, or that its results have no value. Many journals suffer from a serious backlog, and send most papers back unread (this is, properly speaking, not a rejection—for the paper has not even been examined or evaluated); sometimes the editor picks the wrong referee, or a referee with an ax to grind, or a referee who did not understand the paper; sometimes the editor misunderstands the referee's report; sometimes the referee is just plain wrong. Some of my own most influential papers have been treated rather shabbily. I know even Fields Medalists who tell horror stories of papers rejected. One of the secrets to success in the academic game is perseverance. If your paper is accepted the first time around, then congratulations. If not, you should try to be objective and figure out why. Then act intelligently on that new information.

If your paper is accepted, then the referee will most likely have offered comments and suggestions. Some referees go so far as to suggest alternative proofs, different references, or entirely different approaches. Some editors will instruct you to read the referee's remarks, make those changes that you wish, and then to send the final version of the manuscript, labeled "revised" and with a new submission date, to the journal; other editors will explicitly make final acceptance conditional on your responding in detail to everything that the referee has said. In this last case, if you want to continue doing business with the journal (you always have the fallback option of withdrawing the paper), then you are honor bound to respond to *each of the referee's remarks.* The best way to respond is to treat the referee's remarks one by one, and to record in a cover letter to the editor a brief description of just what you did in each instance. In some cases, you may say "the referee is mistaken and here is why." Or you could even say "this is a matter of taste and I respectfully disagree." In most instances you can expect the referee's comments to be accurate and useful and you will probably want to implement them in some form.

If you feel that the referee has been particularly helpful, then you may wish to add a sentence to the paper—alongside your other acknowledgements—saying that you thank the referee for useful suggestions. You will find it awkward to endeavor to determine the identity of the referee, so do not plan to mention the referee by name.

When you are finished with your revision, assuming that a revision is what has been requested, then make the usual number of copies of the revised manuscript, mark each of them "Revised" and put the date, and then return these to the editor along with a new cover letter. Your new cover letter should state plainly that this is a revision of a previously submitted paper, that you have responded to the referee's remarks, and that you consider this to be the final copy. Please note, however, that the editor *might* send the revised paper to the original referee—or to some other referee—again, and you may be asked to make even further changes. You can expect to receive an acknowledgement of your new submission, together with a clear statement of whether this is the end of the road or whether you will be hearing again from a referee.

And now, as the Managing Editor of a journal, I would like to ask you a favor. Do not submit your work to a traditional hard copy journal by using *e*-mail. It requires some effort to download such a submission. Also, there are often problems with compiling said paper because the author forgets to send in his local style macros, or the author is using unusual fonts, or the author is using some strange, or buggy, version of TEX. Many authors do not even realize what the problems are because they have learned TEX on some local university system and do not know the difference between out-of-the-box TEX and TEX as customized by the local gurus at that university. (In fact, I have had the experience of saying to an author "You are not using a standard version of TEX; please send me your style macros and the following fonts." The reply that I received was "What's a macro? What's a font?") In short, electronic submission can cause no end of irritation to your traditional paper journal. The journal expects the author to do a little legwork. So please do it. Send in the requisite number of hard copies. Later, if the journal so desires, it will request your TEX ASCII file *on a diskette*. (Here ASCII stands for American Standard Code for Information Interchange.)

After a suitable number of iterations of the procedures just described, you and the journal will reach some closure. Then you must wait—this wait could be from six months to two years or more—for the galley proofs or page proofs of your paper. These you must proofread meticulously, both for mathematical accuracy and for typesetting accuracy. There also will be "Author Queries", noted by hand, on the proof sheets. You must respond to each of these. You should mark your corrections and responses in *red*.

Learn to use the basic proofreader's marks. These should always be written in the margins; use a caret or a line to indicate where the correction is being made, but indicate precisely what the correction is *in the margin*. Avoid, if you can, inserting arrows all over your proof sheets; these are difficult for the typesetter to decipher. The references [Hig] and [SG] have detailed treatments, with examples, of the use of proofreader's marks.

You will always be asked to turn your proof sheets around rather quickly—often within 48 hours. You will sometimes be asked to sign a statement saying that you approve this version of the manuscript going into print. *And you will be asked at this time how many reprints you want.* Most journals provide 25 or 50 or 100 free reprints; but you are always given the option to purchase more. This will probably be the only time that you will be offered this option, as reprints are printed at the same time as the bound journals (reprints are created from unbound copies of the journal issue). After you send back the proof sheets (be sure to use the mailing label or address enclosed—you will probably *not* be returning the proof sheets to the editor, but rather directly to the typesetter), then your job is done. Just wait for your paper to show up in the library, and for your box of reprints to arrive at your doorstep.

2.8   A Coda on Collaborative Work

I have written a great many collaborative papers, and some collaborative books as well. I know others who have never collaborated. And there are others still who have collaborated a few times and would never do so again. Which characteristics lead to a successful and happy collaboration and which do not?

First, if you agree to collaborate on a project (and both parties had better agree at the outset; do not leave this question until the project is finished!) then set aside all questions of priorities. At the end of the collaborative process, it is both painful and inappropriate for one author to say "Well, you didn't contribute very much. My name should go first" or, worse, "Your name should not appear at all." If, at the end of the first paper, either or both participants deems the collaboration unsatisfactory, then the authors can go their separate ways. But, in my view, an agreement to collaborate is an *a priori* contract to see things through to the end.

Some of my collaborations involve multiple papers; in one case the joint work amounts to thirteen papers and a book. In this monumental collaboration, both my collaborator and I know that on some papers he contributed more and on others I contributed more. I can honestly say that neither of us dwells on the matter. Taken as a whole, we are both quite pleased and proud of the *oeuvre*. As it happens, one of us has lost interest in this subject area and the other one has pushed on a bit further, either writing papers alone or in collaboration. This has worked out well, because each of us respects the other.

And this last point is the real key. I know of collaborations in which one author purposely introduced errors into the joint paper in order to see whether the other author was truly reading the paper or not. I know of a collaboration that got to the stage of the paper being submitted to a journal; after a time the authors had a dispute, and one author unilaterally withdrew the paper and resubmitted it elsewhere under his name alone. I know of a collaboration between two lifelong friends who were developing their twentieth paper together; they could not agree on whether to call the first result Theorem 1A or Theorem A1; the matter ended with lawyers, death threats, and guns brandished in the air (this *really* happened). In all these cases the base of the difficulties was that the authors did not respect each other.

None of the scenarios described in the last paragraph should take place when you engage in a collaborative effort. You should enter a collaboration with the view that this is an adventure and you will each see what you can derive from it. Each of you should respect the other(s), and will take great pains to be courteous and helpful to the other(s). The goal is to produce a nice piece of work—*not* to squabble over credit, *not* to argue over whose name should go first (alphabetical is almost always best), *not* to argue over whether future papers will be joint or will be written separately.

This last point can lead to sticky wickets, even between the most well meaning of participants. Imagine this scenario: Mathematicians A and B write one or more joint papers. The collaboration then seems to go into remission. That is, each author goes his own way, and they have little contact for a couple of years. Then one of the authors (say A) cooks up another idea and writes a new paper, by himself, which in some sense builds on the ideas in the old series of joint papers. Mathematician B gets wind of this new paper, feels that his contribution to the earlier work justifies his name appearing on this new paper, and relates this feeling to A in no uncertain terms. Mathematician A feels that the joint work was long ago and far away. The main reason for the existence of the new paper is *his* new idea—which is due to him alone.

Mathematician A feels that B has already received adequate credit for the joint work; no further credit is due B. As you can imagine, a major fight ensues.

This situation is most unfortunate. Nobody is right and nobody is wrong. Here is what *should* have happened—in the best of all possible worlds. Realizing that this new paper builds on old joint work with B, mathematician A should have phoned B and told him about it and then said "I think it would be appropriate for this new paper to be joint between us. What do you think?" Mathematician B, ever the gentleman, should then have said "Oh no, this is your idea. Write the paper by yourself. You can thank me in the introduction if you like." Having participated in transactions of this nature, I can tell you that this is a most satisfactory way to handle the matter. Typically, mathematician B is not hungry for another paper; he just wants his due. Typically, mathematician A is not anxious to offend B; he just wants credit for his new idea. (Of course the human condition is such that there are always more complex forces at play. Perhaps A feels that, in the world at large, B is generally given more credit for the collaborative work than A. Perhaps B feels that A never pulled his weight in the first place and therefore A owes B. Fortunately, this is not a tract on psychology, so I shall not comment further on these complexities.)

By touching base with your collaborators in a courteous fashion, you can usually avoid friction. And the effort is worth it. To go to a conference and run into a former collaborator is a pleasure. If the relationship is healthy and friendly, then there is plenty to discuss and the potential for future joint work always lies in the offing. If instead there is friction and resentment between you and your former collaborator, then meeting again could be perfectly dreadful. This is another skeleton in your closet. Bend over backwards to avoid such a liability.

Chapter 3
Exposition

Reading maketh a full man, conference a ready man, and writing an exact man.

Francis Bacon
Essays [1625], Of Studies

You can fool all of the people all of the time if the advertising is right and the budget is big enough.

Joseph E. Levine

When Kissinger can get the Nobel Prize, what is there left for satire?

Tom Lehrer

Life should be as simple as possible, but not one bit simpler.

ascribed to Albert Einstein

If you have one strong idea, you can't help repeating it and embroidering it. Sometimes I think that authors should write one novel and then be put in a gas chamber.

John P. Marquand

Writing comes more easily if you have something to say.

Scholem Asch

Considering the multitude of mortals that handle the pen in these days, and can mostly spell, and write without glaring violations of grammar, the question naturally arises: How is it, then, that no work proceeds from them, bearing any stamp of authenticity and permanence; of worth for more than one day?

Thomas Carlyle
Biography (1832)

3.1   What Is Exposition?

Perhaps the highest and purest form of mathematical writing is the research paper. A research paper, in its best incarnation, contributes something useful and insightful to our collective mathematical knowledge. If it is very good, then the contribution may live for a long time. The creation and publication of research is what mathematics is all about.

But our profession involves us in other types of writing. We must write letters of recommendation. We must write referee's reports. We must review cases for tenure and promotion. We write surveys. We sometimes write book reviews. We may be called on to write opinion pieces. The present chapter concentrates on *expository writing*.

In its simplest form, mathematical exposition could take the form of a survey of a field on which you are an expert. Or it could be a text or monograph on some specific area of mathematics. The new challenges present in such a writing task are these: **(i)** you are attempting to reach a broader audience than that which would read one of your research papers; **(ii)** you must strike a balance between how much mathematical detail to give and how much explanation and/or handwaving to provide.

The reader of an expository article does not want to work as hard as the reader of a research article. Envision your reader sitting on a park bench reading your expository article, or putting his feet up and drinking a cup of coffee while reading. *Do not* imagine your reader with a pencil gripped in his fist, slaving away over each detail of your paper. Thus, if you are writing an article about the influence of the Atiyah-Singer Index Theorem on modern mathematics, you certainly will not prove the theorem. To be sure, you will refer to some of the excellent books on the subject. You will explain how the result is a far-reaching generalization of de Rham's theorem and the Riemann-Roch theorem. You will describe the ingredients of the proof, and will give a rough sketch of its structure. But you will certainly not *prove* the theorem.

You also will not assume that your reader already knows all the jargon in the subject. You will not assume the reader to be expert in K-theory or pseudodifferential operators. Nor will you assume that your reader is familiar with the motivation for, and the applications of, the subject. *You should not assume that your reader has the perfect background to read what you are writing.*

So you have your work cut out for you. Expository writing is a lot like teaching. You frequently must anticipate your audience's short-comings, and make suitable adjustments in your presentation. But in expository writing you must be smarter than you are when you are a calculus teacher. In the latter situation, your audience is before you and is sending you signals. When you are writing, your audience is (if you are lucky) only in your head.

3.2 How to Write an Expository Article

For the purposes of this section, the phrase "expository article" means a survey article. Such an article might be a survey of some field on which you are expert. Perhaps you are one of the pre-eminent experts, and therefore the canonical person to be writing this survey. Or your colleagues have called upon you to plant a flag for the subject. One scenario is that you have been invited to give a one-hour talk at a national meeting of the American Mathematical Society. It is natural, and commonplace, for you to turn such a talk into a survey that will appear in the *Bulletin* of the AMS. But such an august occasion is not a necessary condition for the writing of a survey. You may simply feel that a survey is needed, that certain wrongs need to be righted, or that the time is ripe. More and more journals are soliciting expository articles; there is a market for high-level exposition done well.

In order to write a good survey article you will need a detailed outline before you. You will be covering a lot of ground, and you do not have the luxury of hiding behind the details of the proofs. In fact, you will most likely not be presenting any proofs in their entirety. When you do choose to present a proof, it will probably be a sketch, or a pseudo-proof. If you are clever, you will present a well-chosen example, work it through, and then say "this reasoning also shows you how the proof works." Or you might say

> This example is in fact the enemy. The proof shows that this example represents the only thing that could possibly go

wrong, and then systematically shows that the hypotheses rule the example out.

A good survey should build to a climax: Poincaré looked at this example, Lefschetz looked at that example, eventually people realized what was going on, and the Eilenberg-Steenrod axioms were formulated (I am thinking here of the genesis of algebraic topology). Alternatively: first there was the Laplacian, then there was general elliptic theory, then there was the $\overline{\partial}$-Neumann problem, then pseudodifferential operators evolved. Simply to begin citing technical results in chronological sequence is not to write an effective survey. You are telling a story, and you must create a tapestry.

A good survey should have a stirring conclusion. By this I do not mean "That's all, folks!!" or a hearty cry for more and better research on Moufang loops. Instead, your survey should conclude by taking a look back at what ground has been covered and where the subject might go in the future. It should note the historical turning points (which you have, I hope, described in the body of your survey), and make speculations about what the future milestones might be. If appropriate, it should sum up what this subject has taught us so far, and what it might show us in the years ahead.

A good survey should have an extensive bibliography. You are not doing your job if you merely say "The three standard books in the subject are these, and you should look in their bibliographies for all the technical references." By all means mention the three standard treatises, and extol the virtues of their bibliographies. But you *must* create your own bibliography. Your list of references is your detailed definition of what the subject is, what are the most important papers, and what is the latest hot stuff. Compiling a good bibliography is a lot of work (though, with the aid of modern technology, not nearly so arduous as in years past—see Sections 2.6, 5.5). But this effort is a necessary part of the process, and the result will be a valuable tool for both you and your colleagues in the years to come.

Writing a good survey—one to which people will refer for many years—is one of the hardest writing tasks there is. Getting all the basic ideas on paper, and in the proper order, is just the first step. Once that task is completed, then you must craft the piece into a compelling tale with introduction, entanglements, climax, denouement, and finale (much as in a Shakespearean tragedy).

Be absolutely certain that you have not slighted any of the players in the subject, nor inadvertently misrepresented their contributions. Almost certainly, in the course of writing your survey, you will be saying (perhaps *sotto voce*) "Here is the right way to see things (implying, perhaps, that some others are not the right way) and here are the important contributions." Whether you do this consciously or not, you certainly will do it. Take extra care that you are diplomatic, and that you let everyone's voice be heard.

The reader of your survey should come away from it feeling that he has been given an entrée to some new mathematics. He should have **(i)** learned some new facts and **(ii)** seen some new techniques. Forty pages of descriptive prose, without any substance, will not wash with a mathematical audience. You must sketch how the ideas unfold, and endeavor to give some indications of the proofs. When deciding what to prove, you must balance what is instructive against what is feasible in a short space. Often you can prove only a special case, or you might say "to simplify matters, we add some hypotheses." As an instance, proving the inverse function theorem for a C^1 function is hard. But if you assume that the function is C^2, then you can use the remainder in the Taylor expansion to good effect and the proof suddenly becomes easy (see the details in [Kr1]). The key ideas are still present, and they come out much more clearly.

Just as when giving a talk, you can fudge a bit in your survey. State theorems precisely and correctly but, when presenting the proof, say "For simplicity, we consider only a special case" or "For a quick proof, let us assume that the function is actually real analytic." Readers will appreciate being given a nugget of knowledge, without the gory details. A research paper should contain complete proofs—proofs that are categorical, and leave no doubt of their correctness. A survey acts as a pointer to the research literature; it is not usually the final word on the proofs.

3.3   How to Write an Opinion Piece

Most mathematicians agree that writing good exposition is considerably more difficult than writing good mathematics. As has already been described in this book, the latter activity makes few demands on your abilities as a creative writer. You need only exercise some taste in organizing and presenting the ideas. However, when you are expositing, then you are less engaged in statement and proof and more engaged in description, explanation, and opinion formation. There is much more latitude and therefore you, as a writer, must exercise more control.

Let us turn now to the writing of an opinion piece. The writing of an effective opinion piece will involve all the skills already noted as mandatory for good expository writing. But the opinion piece must also, if it is to be effective, have fire and life and drive. It must capture the reader's attention, and it must *convince* him of something. How does one go about pulling this off?

A parody of midwestern political oratory has the would-be congressman declaiming

> Agriculture is important.
> Our rivers are full of fish.
> The future lies ahead.

I hope that, when you write your opinion piece, your thoughts have more focus than this politician's, and your message is more incisive and more substantive.

First, to repeat one of the main themes of this book, if you are going to write an essay expressing an opinion then you must have something to say. And you must know clearly and consciously what that something is, and how you propose to formulate it and to defend it. It does not wash to say, in your mind's voice, "I am going to write an essay in opposition to the teaching of calculus in large lectures because it is a bad idea and I hate it." In point of fact most of us agree with this thesis, but it turns out to be extremely difficult to harness facts and arguments to support the thesis. Couple this lack with the fact that there are articulate and serious people—who spend their professional lives studying such matters—who disagree vehemently with the thesis (see [Dub]) and can marshal forceful arguments against it, and your obvious essay in support of an obvious contention suddenly becomes quite painful.

In fact the statistics that bear on the "large lecture" question are a bit unsettling. They tend to suggest that students taught in small classes feel better about themselves and about the subject matter (than do students taught in large classes); they do *not* tend to suggest that such students will turn in a better performance. That is the trouble with facts: they sometimes force you to conclusions that differ with your intuition.

What I am suggesting in the preceding paragraphs is that the writing of a position paper or an opinion piece often involves considerable research. This is not the same sort of research that one performs in order to prove the Riemann hypothesis. But it needs to be done, and done thoroughly. There is no substitute for knowing what you are talking about.

You must have a deliberate and explicit formulation of your thesis and your contentions. Best is to enunciate that thesis in your first paragraph. The thesis could constitute your first sentence, or it could be the culmination of suitable background palaver that lays the history and orients the reader's mind toward the main point of your essay. But the thesis you are defending should be put forth—so that it cannot be mistaken—at the outset of your opinion piece.

The next (major) portion of your essay should consist of cogent presentation of material gathered in support of the previously enunciated thesis. This prose could include facts, reports of studies, anecdotes, logical arguments, and other materials as well. Note that a defense consisting entirely of anecdotes is at first entertaining but, in the end, not convincing. On the other hand, a defense consisting only of dry facts and logical arguments does not generally hold the reader's attention and is not forceful. (The validity of this last statement depends, of course, on your audience, on your subject matter, and on the context. Obviously, most any mathematical research paper contains just facts and logical arguments. But such a recondite exercise is directed toward specialized researchers with an *a priori* interest in what the writer has to say. The audience for an expository paper or opinion piece is more diffuse, less well prepared, and less patient.)

The last portion of your position paper should sum up the major points you have made, repeat the most important ideas, and force the desired conclusion. These are the final thoughts with which you will leave the reader. They are analogous to the closing arguments in a jury trial. Weigh each word carefully. Remind the reader what he has read, and why.

I do not mean to suggest that persuasive writing is formulaic. This activity is not like laying bricks. Some of the best position papers conform only loosely, if at all, to the rubric just laid out. But the points I have made, and the issues I have raised, are salient to any polemic, no matter what its exact form.

3.4   The Spirit of the Preface

Many writers spend little time in writing a preface; some forget to write it at all. This is a mistake. Your prefatory remarks are often the most important part of your writing. They tell the reader why you write what you write, what your goals are, and what you intend to accomplish. They state what you assume, and what you conclude. These principles apply whether you are writing a book (which has a separate, formal preface) or an article (which may have a prefatory section, or collection of paragraphs) or a letter (which may have just one prefatory paragraph). The preface is your statement of purpose. It is vital to your mission.

When an editor at a publishing house receives a mathematical manuscript (for a book, say) for consideration, he usually seeks advice from one or more experts in the field. When I am asked by a publisher to review such a manuscript, the first thing I look for is a Preface or Prospectus (a marketing version of the Preface), and a Table of Contents. These two items, if present, will give me a quick overview of the project: What material is covered? At what level? Who constitutes the intended audience? What are the prerequisites? What need does this book fill? What are the book's competitors? (In modified form, these queries also apply to an expository article. If I receive a 100-page expository article to review, then I hope that it contains the information that a book Preface and Table of Contents usually provide.)

Without this information, I have no idea what I am reading. The manuscript could start out with sophomore-level differential equations, and before long be doing canonical transformations for Fourier integral operators. As a result, I have no idea what the author is trying to accomplish.

Whether you are writing a research announcement, a research paper, an expository paper, a book, or virtually anything of a scholarly nature, you should always ask yourself the questions in the second paragraph in this section. Most importantly, you must decide *in advance* the book's intended audience and you must, at all times, keep that group clearly in focus. If you are writing a calculus book, then presumably the audience is freshmen, and therefore you must resist the temptation to indulge in asides to the professor. If you are writing a research article, then presumably the audience is fellow researchers, like yourself—not Gauss and God. If I may be permitted a little hyperbole, I will say now that having a strong sense of your audience is the single most important attribute of an effective writer.

I have sung the praises of the Preface. But the Table of Contents (known as the TOC in the publishing industry) is nearly as important. Here is why. When you are writing a book, which is a big project, you should have the entire scope of your endeavor firmly planted before your mind's eye. In this way you can measure your task, you can see what progress you are making, and you can keep the affair in perspective. Sometimes you can have fun just sitting down and starting to write, or just seeing where your thoughts will lead you, or modifying your project every time an interesting new preprint comes across your desk. But let me assure you that these methods are a sure way to guarantee that your book will never be completed. Writing the TOC addresses this impasse.

Now you may not be interested in writing something like a book. Such a writing project is awesome and onerous; the task is not for everyone. But the principles in the last paragraph apply even to writing a twenty-five-page research paper. You need to have the full scope of the paper in your mind so that you can endow your working methods with a pace and give yourself a sense of incremental accomplishment. This sort of organization is also just a simple device for keeping yourself from becoming completely confused.

I often begin a book by writing the Preface, because it helps me to organize my thoughts and to orient myself toward the project. I refer to it frequently as my work on the project progresses. At the same time, it also makes sense to write the Preface last: for when the book is complete, then you know in detail what you have written and you can describe it lovingly to the reader. My recommendation is to do both. Write a version of the Preface before you begin the book. When you are finished, write it again.

This section on the Preface may seem like a digression, but it is not. Even if you are writing an opinion piece (Section 3.3), or a letter of recommendation (Section 4.1), or a book review (Section 4.2), your piece should contain prefatory remarks. Such remarks are good for your reader, so he knows what this piece of writing is supposed to be about. But, most importantly, they are good for you: they keep you honest, and keep you on your course.

3.5   How Important Is Exposition?

There are those who will argue that mathematical exposition is not important at all. The one true pursuit, whose fruits are recognizable and of lasting value, is mathematical research. Some would go so far as to claim that even the writing and publishing of research results is an activity suitable only for hacks. Instead, for a good mathematician, it suffices to prove the theorems, tell at least one other person about them, and then let the word spread.

I happen to think that the attitudes described in the preceding paragraph are counterproductive. Scholars are not monks. They are an active and engaged component of society. Along with the universities and research institutes, they are the vessels in which mankind stores its accumulated knowledge and civilization. Thus scholars must communicate. They can do so by giving lectures; the importance of lectures cannot be over emphasized. But scholars also must write. The written word—unlike the spoken word—lasts for ages, and can influence generations.

Exposition is important because it reaches a broader audience than do specialized research articles. Thus good expository articles disseminate information quickly, and they are much more likely to spawn collaboration between different fields than are specialized articles. An outstanding expository article will cause even the experts to reorganize the subject in their own minds.

In my own work, I have found that expository writing is a device for teaching myself. It forces me to organize my thoughts, and to be sure that I understand how a subject is constructed—from the ground up. This is also a device for teaching my students: after I have explained the same idea several times to several different students (perhaps over a period of years), I find it useful to write something down. Then, when the next student comes along, I can give him something to keep.

I can still remember, many years ago, reading an article by Freeman Dyson called "Missed Opportunities" [Dys]. I have never seen anything like it before or since. In this article, the author makes statements like the following: "In 1956, X proved this and in 1957, Y discovered that. If only I had been alert, I could have combined these results with ideas of my own and with Heisenberg's uncertainty principle and I could have done thus and such. Instead, Z combined the ideas in a different way; for this work he later won the Nobel Prize."

I found this article to be an inspiration in several respects. First, I was amazed that anyone could understand his subject so well that he could recombine its parts in ways that never actually occurred. Second, I was amazed by Dyson's candor. Third, Dyson helped me see what creativity is. Fourth, he gave me a sense of the scope of knowledge.

Dyson's article is a sterling example of what good exposition can do. In a lifetime, you probably will not read more than half a dozen articles that are this good. But half a dozen is enough. If you are truly fortunate, and extremely talented, then perhaps you will write one.

Chapter 4
Other Types of Writing

Stand firm in your refusal to remain conscious during algebra. In real life, I assure you, there is no such thing as algebra.

Fran Lebowitz

Neither can his Mind be thought to be in Tune, whose words do jarre; nor his reason in frame, whose sentence is preposterous.

Ben Jonson
Explorata—Timber,
or Discoveries Made upon Men and Matters

The flabby wine-skin of his brain
Yields to some pathological strain,
And voids from its unstored abysm
The driblet of an aphorism

The Mad Philosopher, 1697
in *The Devil's Dictionary*
by Ambrose Bierce

What is written without effort is in general read without pleasure.

Samuel Johnson

Sometimes a cigar is only a cigar.

Sigmund Freud

The good writing of any age has always been the product of someone's neurosis, and we'd have a mighty dull literature if all the writers that came along were a bunch of happy chuckleheads.

William Styron
interview, Writers at Work (1958)

Close your eyes and think of England.

—a Victorian mother, giving advice to her daughter
concerning behavior on the wedding night.

4.1   The Letter of Recommendation

Once you have become an established mathematician, you are likely to be asked for letters of recommendation. Such a document could be a letter of recommendation for a tenure case, or for a promotion, or for both. It could be a letter recommending a young person for a first or second job. It could be a letter recommending a senior person for an endowed Chair Professorship, or for the Chairmanship of a department. (For the sake of this discussion, I will call these "professional letters".) It also could be a letter of recommendation for a student (such letters are treated a bit differently from professional letters—see below). There are many variants; here I would like to distill out some unifying principles on writing letters of recommendation.

When you are asked to write a letter as described in the first paragraph, you are in effect set a task. You have become a one person "taskforce". What makes a taskforce different from a committee is that a taskforce is not supposed to debate the task at hand; instead, the taskforce is supposed to perform the designated task. In the present instance, you are supposed to offer your professional opinion on a certain matter.

In my view, it is both unprofessional and irresponsible to dodge the assigned task. Let me be more precise. There certainly will arise circumstances where you either cannot write or should not write. Perhaps you have had a fight with the candidate in question and feel that you cannot offer an objective opinion; perhaps you have a conflict of interest; perhaps you are unfamiliar with the general area in which the candidate works; or perhaps you do not know the candidate well at all. In any of these cases, or similar ones, you should quickly and plainly write to the person (the dean or chairman) who requested the letter and say that you cannot write. Best is if you can give the reason, but it is acceptable if you cannot. Do not agonize over the task for six months and *then* decline to write; take care of the matter right away.

The circumstances described in the last paragraph should be considered to be extreme exceptions. They will come up less than one percent of the time. In most instances, you will be asked to write about some particular person for some particular circumstance and you should say "yes" and then you should do it.

I know mathematicians who will agree to write an important letter and then not do it. This paradox usually occurs for one of two reasons: **(1)** the putative letter writer is pathologically disorganized and forgets, or **(2)** the putative letter writer has nothing nice to say about the issue or person at hand and does not want to say it. I have already addressed the second of these *conundra*. The first of these situations is not likely to arise if the request to write was submitted to you as a formal letter—from a dean, for instance. For then the piece of paper is sitting somewhere on your desk and you will probably get to it eventually. The paradox *can* occur if instead a student pokes his head in your door and asks for a recommendation to graduate school. You give a cheery "yes" and then the entire matter vacates your head. To avoid this error, ask the student to put the following on a slip of paper: his name, any classes he took from you or other pertinent data, and the address to which the letter is to be sent. (This ruse also helps you to avoid the embarrassment of having to ask the student's name.) Now you have it in writing. Also ask the student to come back in a week or so and remind you. I usually find it convenient to write the letter right away (if the student has poked his head in the door then it is likely my office hour and I might as well be doing *something*). For once the request has been tendered, I am probably already thinking about what I am going to say; I might as well write it down and be done with it.

Make a point of writing requested letters in a timely fashion. It is the professional thing to do, and you would appreciate such consideration if the letter were about you.

Having decided to perform the task—that is, to write the requested letter—you must do what you have been asked to do. That is, you must formulate an opinion, state it clearly, and defend it. The standard format will be explained below.

In the first few sentences, state plainly the question that you are addressing. For example:

> The purpose of this letter is to support the tenure and promotion of Zoltan A. Beelzebub. I have known the candidate and his work for a period of six years, and have been impressed with his originality and his productivity. I indeed

think that tenure and promotion are appropriate. My detailed remarks follow.

Alternatively:

> You have asked for my opinion on the tenure, and promotion to Associate Professor, of Dr. Aloysius K. Foofnar. Dr. Foofnar is now six years from the Ph.D., and in that time has produced nothing but some rotten teaching evaluations and a letter to the editor of the *Two-Year College Math Journal.* Based on that track record, my opinion is that he is worthy of neither tenure nor of promotion.

The bulk, or body, of the letter follows, and it should support in detail the thesis enunciated in the first paragraph. I shall comment below on what might constitute that support. First, let me conclude these beginning thoughts.

Once the body of the letter is written—and this could comprise one or two (or even more) pages—then you must write a concluding paragraph. You *must* write it. You must sum up the point you have made, and restate your thesis. A sample of this practice is

> In view of the stature of Laszlo Toth in the field of computational algebraic geometry, and considering his accomplishments as a teacher and as a scholar, I can recommend him without reservation for promotion and tenure in your department.

(I am assuming that you have in fact described Toth's status and accomplishments, in a favorable manner, in the preceding paragraphs.) Another possibility is:

> In sum, I feel strongly that Seymour Schlobodkin should not be promoted or tenured. Indeed, I cannot imagine the circumstances in which such a move could be considered appropriate.

There are those who, although experienced letter writers, do not adhere to the general scheme just described. One of the standard rationales for this behavior is that, in many states and at many institutions, it is (theoretically) possible for the candidate to have access to the complete text of his letters of recommendation—including the identities of the writers. If such is the case, then the soliciting school will inform the writer at the time the letter is solicited. Of course the letter writer is offered the option up front of declining to write if he is uncomfortable with this "freedom of information" situation.

There are those who, still uncomfortable, agree to write but are afraid to say anything. The most negative thing that they are willing to do is to "damn with faint praise." Not only does this artifice undercut the responsibility of the letter writer, but it puts on those evaluating the case the onus of trying to figure out what the writer was trying to (but did not) say. In the best of all possible circumstances, someone at the soliciting institution will phone the letter writer and just *ask* what the letter was meant to say. In the worst of circumstances, the evaluators are left to guess what was meant. Given that someone's life and career are in the balance here, it is a genuine shame for such a circumstance to come to pass.

Enough preaching. I will now give some advice about the body of the letter. If you want your (professional) letter to have some impact, and to be taken seriously, then you must do two things: **(i)** make some specific comments about specific work or specific papers of the candidate, **(ii)** make binary comparisons. You may also wish to discuss other qualifications of the candidate. No matter what these may be, you should heed these principles: be *precise*, speak of *particular* attributes, and speak only of those topics of which you have *direct knowledge*. Now let me explain.

Your letter had better say more than "Ahmenhotep Smith is a hail fellow, well met. Give him whatever he wants." First, such a letter does not say anything. Second, given the circumstances described above, in which some letter writers attempt to avoid litigation by "damning with faint praise," such a vague letter could be construed as *sotto voce* damnation. If your comments are instead detailed and specific then it is difficult for people to misconstrue them.

Thus you should dwell, for a page or more, on specific virtues of the candidate's scientific work. Make detailed remarks about specific papers: Why is this result important? How does it improve on earlier work? How does the work advance the field? Who else has worked on this problem? This material should not be a self-serving introspection. Remember that most of the readers of the letter will be nonspecialists. Many, including the dean and members of his committee, will not even be mathematicians. Thus attempt, briefly, to give background and motivation. Drop some names. For example, say that Ignatz of MIT worked on this problem for years and obtained only feeble partial results. The candidate under review murdered the problem. If appropriate, point out that the candidate publishes in the *Annals* and *Inventiones*—and that these are eminent, carefully refereed journals.

It is astonishing, but true, that even highly placed people, who write dozens of influential letters every year, seem to be unaware of the need for binary comparisons. To put it bluntly, an important letter that is to have a strong effect *must* compare the candidate being discussed to other people, of a similar age and career level, at other institutions. The comparison should be with people—preferably other academic mathematicians—whose names the informed reader will recognize. Thus, if the candidate is an algebraic geometer and you say in your letter that "this candidate is comparable to Mumford when Mumford was the same age" then most algebraic geometers will know exactly what you mean and will be extremely impressed; they will in turn explain to their colleagues the significance of your remarks. If instead you say "this candidate is comparable to Prince Charles when Charlie was a student at Gordonstoun" then nobody will know what you are talking about—and you can be sure that they will not be impressed.

To come to the point, if you are writing an important letter that you want people to notice, then you must say something like

> The five best people under the age of 35 in this area are A, B, C, D, E.

In the best of all circumstances, the candidate under consideration in your letter is one of A through E—and you should point out that fact. Alternatively, you could say

> Two of the best people in this field, at the beginning tenure
> stage, are Jones and Schmones. Candidate Bones fits com-
> fortably between them. Bones is surely more original than
> Schmones and more powerful than Jones.

Or you could say that the candidate falls into the next group. Or that
the candidate is so good that it would be silly to compare him to the
usual five best. Say what you think is appropriate. But *say something.*
If you do not, then the readers will notice the omission and infer that,
between the lines, you are saying that this guy is not any good. Better
to say that he is number 15 than to say nothing at all.[1]

Tailor your binary comparisons to the circumstances. It would be
inappropriate to compare a candidate two years from the Ph.D. with
a sixty-year-old member of the National Academy of Sciences (unless
the comparison is favorable, and you are trying to knock the reader's
eyes out). It would probably be inappropriate to compare *anyone* with
Gauss (although I *have* seen favorable comparisons with Gauss!). Note
also that, if you are recommending a senior person for (just as an instance) an honorary degree, then binary comparisons might be entirely
out of place, and uncomfortable as well. If the person is already a
Chair Professor at Harvard, then to whom will you compare him? And
to what end?

Your letter of recommendation can contain other specifics and details that might grab the reader's attention. You could say that the
candidate gives excellent talks at conferences. You could say that he
is a wonderful collaborator. You could say that the candidate has
beautiful insights, and that talking mathematics with this person is a

---

[1]A *caveat* is in order if the letter that you are asked to write is *not* solicited
from a research institution. If the candidate is in fact at a four-year college, where
the primary faculty activity is teaching, then the school probably demands a lot
of classroom activity—and not so much scholarship. These days, almost every
school wants its permanent faculty to have some sort of academic profile; but a
teaching college can hardly expect its instructors to stand up to hard-nosed binary
comparisons. The lesson is this: read the soliciting letter carefully; speak to people
in *their language*, and tell them what *they* want to know. If the soliciting letter is
from a teaching institution, then it is probably most appropriate for you to write
about teaching, curriculum, publications in the *Monthly*, and letters to the editor of
UME Trends. A disquisition on Gelfand-Fuks cohomology is probably less apropos.

pleasure. You could describe in glowing and heartfelt terms the process of proving a theorem, or of writing a paper, with the candidate.

These days, credible evidence that the candidate is a good teacher will certainly help the case. Of course you are probably not in the same department as the candidate, so you very well may not be able to discuss his teaching. If the candidate is a truly outstanding teacher, then perhaps you have heard his colleagues mention his talents, or perhaps you know that he has won a teaching award. It makes quite an impression on letter readers if Professor A, from University X, can comment knowledgeably and in detail on the teaching of Professor B from College Y.

Here are some travesties that I have seen (all too frequently) in letters of recommendation. You should certainly not emulate any of these mistakes:

1. The writer begins in one of the fashions indicated above. Then he says

   > Nefertiti Prim has proved the following theorem about pseudographs (state the theorem). This is a nice result. The theorem is based on some old ideas of mine. [*And the rest of the letter consists of a description of the letter writer's own work!*]

   Such a letter violates all the precepts laid out above, and marks the writer as a thoroughly self-absorbed fool. Of course this letter does nothing to help, nor to hurt, the candidate; but it gives a rather poor impression.

2. The writer discusses the candidate, discusses the candidate's work, makes binary comparisons, and mentions specific papers. In short, the writer makes all the right moves. In the concluding paragraph, he says

   > I am going to make no specific recommendation as to whether you should promote Mergetroyd Plotz or not. After all, you know what the needs of your department and your school are. You can use

the information that I have provided to come to an appropriate decision.

Rubbish! Imagine taking your car to a mechanic and hearing him say "Your transmission runs at half speed and your rear wheels turn forward. Your stroke is short and your valves rattle. I am not going to make any specific recommendation for repairs because, after all, it is your car and you know what your needs are." Or imagine your physician saying "Your heart will give out any day now, and you are also a prime candidate for a stroke or for total paralysis. However, I will make no specific recommendations. It is your body, and you know best ...". *You are a professional; you are expected to render an opinion.*

3. The writer neglects to address explicitly the question at hand. This omission is sometimes committed inadvertently, but this omission is a dreadful error. If you are asked whether Hyapatia Lee should be tenured, or promoted, or given a certain post, or a grant, then you must say point blank what your opinion is *about that question as it applies to that candidate.* If you neglect to say, then your letter (taken as a whole) is likely to be read as the worst sort of "damning with faint praise". Whether you intended it or not, you may have buried the candidate.

4. The writer faces the following request (and blows it): In a school that fancies that it wants to make hard decisions, and elicit the *bona fide* truth from the letter writers, it is common for the dean to include in his solicitation letter a query like "Would you tenure Marilyn Chambers in *your* department?" If the person being asked for the letter works at Harvard, and if the institution soliciting the letter is a four-year teaching college, then such a dean is just looking for trouble.

Even if the letter of solicitation does not explicitly ask this question, we letter writers are often tempted to answer it. Unfortunately, the answer sometimes comes out like this:

(∗) Dr. Morris Fischbein is not good enough for us,
but he is certainly good enough for you.

Rarely is a letter writer clumsy enough to phrase things quite this bluntly, but I have seen many a letter in which this sentiment comes through loud and clear.

This is just too bad. The person writing such a statement (or a euphemistic paraphrase of it) probably thinks that he is being frank and helpful. He is being neither. Instead, he is insulting the maximum number of people in the least constructive possible fashion. A word to the wise should be sufficient: proofread your letter of recommendation to be sure that you have not inadvertently (or intentionally) made statement (∗). The inclusion of such an assertion in your letter will vex the readers, and render your letter ineffectual, so that it will not count. I presume that this effect is not the one that you want.

If in fact you are at a place like Harvard, and if your letter is solicited from a much more humble institution, and if you *must* address this difficult question, then you should endeavor to tell the truth. Say that Harvard's math department is usually ranked in the top three; you only tenure people who are world leaders, indeed great historical figures; such standards would be inappropriate to apply at an institution like the one which has solicited the present letter. However, you certainly would recommend this candidate for tenure at Bryn Mawr, or Swarthmore, or some other institution that you choose for comparison.

That concludes my enumeration of woeful errors. Now let me cut to the chase. When you are writing a letter for a candidate, then a heavy responsibility rests on your shoulders. The dean or chairman who solicits the letters of recommendation is not simply casting his net and taking a vote: this person wants a *mandate*. He will *not* weigh good letters against bad: he wants to be socked between the eyes. A tough dean once told me "If a case is not overwhelming then I turn it down. If the candidate is any good, he'll land on his feet. If not, then we are better off without him." Thus if your letter says

> Herkimer Nixon is no good. Don't do it.

then you may as well face the music and realize that *your letter alone* will have killed the case—at least for now. I cannot repeat this point too strongly: it is dead wrong to say to yourself "This is a negative letter that I am writing, but it will not count unless all the other letters are negative too." Baloney! One negative letter will usually stop the case cold. That is all there is to it.

A letter with inadvertent errors (of the sort mentioned above) will not necessarily bury the candidate, but it certainly will not help him.

In the closing paragraph of your letter, you will typically indicate a degree of enthusiasm for the case at hand. Here is a graded list of examples—taken from letters that I have actually seen:

> Igor Stravinsky has done a workmanlike job with his research program.

> Ayatollah Bono is a reasonable case for tenure. You would not go wrong to tenure him.

> I recommend Zigamar Pistachini warmly for tenure and promotion.

> I recommend Rufus P. Quackenbush enthusiastically for tenure and promotion.

> The case for Guy de Maupassant Rabinowitz is overwhelming. I recommend him without reservation.

> I give Chicken à la King my strongest possible recommendation. Phone me if you require further details on the case.

In case my admonitions have not sunk in, let me beat you over the head with them. The first two of these statements are in fact negative. Whether they were *intended* to be negative, or are simply an articulation of the writer's loss for words, that is how they will be read. You might as well take the candidate out and shoot him. The third passage is a little better (many evaluators will read "warmly" as "lukewarmly"), but does not convey passionate affirmation.

By contrast, the fourth example will definitely be construed favorably. The adverb "enthusiastically" conveys the positive nature of the assertion. The last two sample sentences represent the sort of forcefulness that is virtually mandatory if you want to argue for the tenuring of a candidate at any of the best institutions.

The writing of letters of recommendation is not formulaic. Indeed, if all letters of recommendation fit a pattern and sounded the same, or if all *your* letters look the same, then they will eventually be ignored. Mathematicians keep a mental database on letter writers in the same way that good baseball pitchers keep a database on batters. After several years, we know who "tells it like it is" in his letters, who spins tales, and who simply cannot be trusted. We know who always writes the same letter for everyone. And we act accordingly. (In fact there is an eminent mathematician who has had many students and writes a great many letters of recommendation. They are so similar that you could hold any two of them up to the light, one behind the other, and most of the words would line up. But then he scribbles his real opinion in the margin by hand.)

You will develop your own style of writing letters. Mathematics is a sufficiently small world that, after several years, people will recognize your letters of recommendation at a glance. But, no matter how you write your letters, you will want to take into account the issues raised in this section.

During times when jobs are hard to come by, letters of recommendation tend to become more and more inflated. Everyone feels that he must try harder if he is going to land a job for that special someone. Here are examples of lines that have been used to describe specific, rather famous, job candidates. I do not necessarily recommend that you use any of them; if you do, the readers might think that you are eating with only one chopstick. But these examples will give you an idea of what some people have done to draw attention to what they are saying, or to remove their particular letter from the ranks of the humdrum. (Of course names have been changed to protect the innocent.)

Beef E. Wellington has a good idea every other day and writes a brilliant paper every week.

Potatoes au Gratin knows both classical analysis and modern analysis. He is the natural successor to Hardy and Littlewood.

Talking to Leon Czolgosz is like talking to Enriques. (An inspiring thought, written by one too young to have ever spoken with Enriques.)

Cherries Jubilee is the most mathematically intelligent person that I have ever met.

Rootie N. Kazootie is the greatest mathematician since Gauss.

Although there is an art to writing a "professional letter", it is also the case that at least you are dealing with familiar territory, and speaking of matters on which you are expert. Any professional mathematician for whom you might write has a publication list, and a track record in teaching, and a reputation as a lecturer, and some *gestalt* as a collaborator. When you are writing for a student, by contrast, matters are more nebulous. The student has none of the professional attributes that you are comfortable discussing. Yet, if you want your letter to be memorable, and to be perceived positively, then you still want to say something noteworthy about the student.

While the precepts of organization that I have stated above still apply in a letter for a student, some of the other particulars do not. For example, you most likely cannot remark on the student's scientific work, and you most likely cannot make binary comparisons. In fact any attempt at binary comparison is likely to be ludicrous. Imagine saying "I am delighted to recommend Sacajawea Smith. She is every bit as good as Euthanasia Jones, whom I recommended five years ago to a different institution." If in fact you previously recommended a student who turned out to be a well-known star—or at least a well-known star at the institution to which your letter is addressed—then by all means make a binary comparison involving that person if such a comparison is appropriate. Usually it is not appropriate, so no such comparison should be included.

Thus in practice you must try a bit harder to say something specific about the student for whom you are writing a letter. After you have been teaching for several years, it may be the case that you have actually taught a few thousand students (this would be true, for example, if you have taught calculus in large lectures several times). It becomes difficult to distinguish students—even good ones—in your memory, much less to say something of interest about any of them. If you apply yourself to the task, then you can nevertheless come up with some noticeable things to say. Here are some examples, taken from genuine letters:

> Fig Newton is one of the five most talented undergraduates that I have encountered in twenty years of teaching.

> Iphiginea Mandelbrookski is hard working and perseverant. She can think on her feet—at the blackboard—just like a mathematician. She is original and imaginative.

> In order to test her creative abilities, I have given Georgina Spelvin extra work outside of class. She discovered a new proof of Gronwall's inequality, discovered Euler's equation in the calculus of variations on her own, and has also posed numerous interesting problems of her own creation. Needless to say, she breezed through all the standard class work.

As usual, the point is to say *something*—and that something should be quite specific. The view of letter *readers* is that if the letter writers cannot say anything unambiguous and remarkable about a student, then there is probably nothing remarkable about that student. So what if the student can earn mostly *A*'s in his classes?—that is no big deal, and in any event can be gleaned from the transcript.

Sometimes a student, or someone else, will ask you for a letter about himself and you do not feel that you can write a good one. Either you have nothing to say, or you have nothing good to say, or you have some other valid reason for not writing. (Note that this case is different from the one in which a dean is asking you for a letter about one of his faculty. Now the candidate himself is standing before you and asking for a letter *about himself*.) You always have the option of agreeing to write, and then writing a negative letter. Often, however, you bear the candidate no malice and think that he deserves a chance. In that case, the honorable thing to do is to say to the candidate "I'm sorry. I frankly do not feel that I could write a good (or supportive) letter for you. Perhaps you should ask someone else." The rotten thing to do—and this happens far too often—is to say "Oh yes, fine" but with no intention of ever writing *anything*. Note that the lack of your letter in the dossier will make that dossier incomplete; in many cases the candidate will not, as a result, be considered at all. If that is the effect you want, then you should have the courage to say something in a letter. If it is not the effect you want, then you should have the courtesy to take a "pass".

One of the most critical, and delicate, types of letter that you will have to write is a letter seeking a job for a student completing his M.A. or Ph.D. under your direction. Your statements are *a priori* suspect because you obviously have a vested interest in finding this student a job, and in seeing him succeed. Thus you must strive to put into practice the precepts described above: **(i)** say why this student is good, **(ii)** say what this student has accomplished, **(iii)** if possible, compare the student favorably with other recent degree holders, **(iv)** say something about the student's ability to teach, **(v)** say something about the student's ability to interact with other mathematicians.

A meat-and-potatoes job application from a fresh Ph.D. has a detailed letter from the thesis advisor that conforms, at least in spirit, to the suggestions just adumbrated. This detailed letter is accompanied by two or three additional letters from other instructors at the same institution, each of which is rather vague and says in effect "Doo dah, doo dah; see the letter by the thesis advisor." If you want your student's dossier to stand out, and to really garner attention, then you should strive to help the student make his dossier rise above this rather dreary norm. Endeavor to ensure that the other writers know something about what is in the thesis. If possible, convince someone from another institution to write a letter for the student. Make sure that the dossier includes detailed letters about the student's teaching abilities.

When you write a letter of recommendation, tell the truth. If all your letters read "This candidate is peachy, and a dandy teacher too. Give him X" (where X is the plum that the candidate is applying for), then after a while nobody will pay any attention to what you say. I presume that if you take the trouble to write letters, then that is not the result that you wish. The infrastructure has a memory. It will remember whether you are a person who can make tough decisions, or whether you are wishy-washy. If you want your letters to count, then you must call it as you see it. It is hard to be hard, but that is what the situation demands.

One issue that we, as letter writers, often must address is whether or not a job candidate can speak English, and how well (this question could even apply to an undergraduate student—especially if that student is applying to graduate school and might be considered for a Teaching Assistantship). In this matter we are, in the United States, cursed by our group dishonesty over the past twenty years. Too often have we said in a letter that "this candidate speaks excellent English, can teach well, and is a charming conversationalist to boot." In a more frank mode, we might have said "This candidate speaks better than average English" (recalling Garrison Keillor's statement about the town of Lake Woebegon, in which "all the children are above average"). When the candidate arrived to assume his position, the hiring institution often found that he could not understand even simple instructions and had no idea how to teach.

It is difficult, but you must endeavor to be honest about the candidate's fluency in English (again, your credibility—which will follow you around all your life—is at stake). You could say, for example,

> This candidate speaks English like a home-grown American, with no trace of an accent. Listening to him is like listening to Walter Cronkite.

This would be the ideal thing to write, and would dispel all trepidations about the candidate's fluency. Unfortunately, if the question needs to be addressed at all, then this statement probably is not true. You could instead say

> Luisa Longshoremanska has been taking "English as a second language" and has taught several lower-division courses successfully. Her English is accurately formulated and clearly enunciated. Students have no trouble understanding her.

Unfortunately, you cannot always be so enthusiastic. Sometimes you must say something like

> Mr. Anthrax Xlpltqlpl has been working hard on his English, and has made substantial progress. One still needs to concentrate in order to understand him.

Or you might say

> It takes students three or four days to become accustomed to Ms. Imelda Rasputin's English, but her charming personality helps them along. As a result she is a most successful teacher.

The thought that I am trying to formulate here is that Mr. Xlpltqlpl's English or Ms. Rasputin's English is not perfect. But Mr. X and Ms. R are real troupers. They try hard, and the students (at least in Ms. R's case) forgive them a lot.

Of course you can plainly see that I am trying to suggest ways to avoid saying "This person cannot speak English and refuses to learn. He is only suitable for a nonteaching position." But sometimes—presumably not in the case of Mr. Xlpltqlpl or Ms. Rasputin—it must be said.

At the risk of repeating myself, let me say that when you address the candidate's ability with English then you should not be formulaic. If all your letters about foreign candidates say

> *X*'s English is just fine. He is a good teacher.

then, after a while, the world will mod out by that portion of your letters. Try to say something original, apt, and true about each candidate. I once wrote the following about a fresh Ph.D., from a foreign country, who was applying for a job:

I consider myself to be rather a good teacher, but I really learned sómething when watching Mr. Frangi Pani with his class. He moves skillfully among his students, looks at their work, makes insightful remarks, and does a marvelous job of eliciting class participation. It is clear that the students like and respect him.

This passage addresses the language issue implicitly, for it confirms that the candidate can *teach*. Moreover, it is not just a bunch of pap. It says something particular and notable about the candidate's abilities.

Occasionally, you will have to address a truly thorny matter in one of your letters of recommendation. As an instance, I was once writing on behalf of a young mathematician who was applying to several dozen first class universities for a position. I thought that I knew this person quite well. But, a few days before I was going to draft my letter, I learned that the candidate was undergoing a sex change. I had to decide whether I should mention this fact in my letter. I reasoned as follows: if he were changing from Catholicism to Judaism, or from Democrat to Republican, or from carnivore to vegetarian, I certainly would not consider discussing the matter in my letter of recommendation for a mathematical post; so why should I treat trans-sexuality? And I did not. Some time later, I discussed the matter with one of my mentors. He told me that I had erred. In stern terms, he informed me that a matter like this could affect the candidate's ability to teach, and his ability to function as a colleague; therefore I was morally obligated to mention the matter. I still do not know what the correct course of action should have been. I only hope that I will not be faced with another choice like this one any time soon.

Just for fun, let me conclude this long section by quoting from a letter for tenure that was written (truly!) about twenty years ago for a candidate in a French department. Call the candidate Mr. de Gaulle.

Surely Mr. de Gaulle is now wiser than he once was.

That was the entire text of the letter!—No introduction, no conclusion, no binary comparison, no exegesis of the candidate's scholarly work. Just the one sentence. Although the letter does not follow the precepts described in this section, it definitely gets its point across.

4.2 The Book Review

As with most topics in the subject of writing, there is some disagreement over what constitutes a good book review. When Paul Halmos was the book reviews editor of the *Bulletin* of the AMS, he sent every reviewer a set of instructions. The gist of these instructions was that a book review is not a book report. It should *not* say "Chapter 1 says this, while Chapter 2 says that. Chapter 3 is a bore, and Chapter 4 is too hard."

Instead, according to Halmos, a book review on a book about X is an excuse to write an essay about the subject X. Look at the book reviews in the *New York Review of Books*. On the whole they are a delight to read, and they conform to Halmos's view of what a book review is and does. These reviews tell you about the book, but they paint the picture on a large canvas.

To reiterate: If you are reviewing a book on harmonic analysis, then you should write about the history of the subject, what the milestone books and theorems have been, who the major players are and were, and what the big problems are. Drop some names. Make some assertions and conjectures. Having laid considerable groundwork, then finally focus on the book under review. Describe where it fits into the infrastructure you have outlined. Indicate its strengths and weaknesses. Suggest who would profit from reading it, and why. Touch on areas where there is room for improvement. Do not, however, use my suggestions here as an excuse to write an opinionated essay and virtually ignore the book. The book review is supposed to be *about the book;* but it should be about the book in the context of the subject matter, not the other way around.

Here are some other issues that your book review might address:

- Will students benefit from reading the book?

- Are there exercises?

- Are there lists of open problems?

- Is there an accurate and complete bibliography?

- Is there an index?

- Is there a list of notation?

- Is there sufficient review material? Does the book begin at a reasonable level?

- Does the author provide an adequate amount of detail in the book? Does the book make too many demands on the reader?

- Are the proofs complete, clear, and accurate?

- Is the book organized in an intelligent fashion that is useful to the reader? Can the beginner navigate his way through the book?

- Is the history correct? Are attributions complete and accurate?

- Does the book bring the reader up to the cutting edge of research?

If you think about the issues that I have raised here, then you will realize that I have described what a potential reader of the book will want to know when he is making a decision as to whether to buy the book and whether to read the book. One of the main purposes of your review is to inform such decisions.

Many mathematical book reviewers—writers for the *Bulletin* of the AMS, for instance—feel obligated to write a *positive* or upbeat book review, no matter what they really think of the book. They are afraid to be critical. In my opinion, this attitude is an error. Not all books are good, and not all good books are entirely good. You will help the audience, and the author as well, if your review points out inadequate features of the book, or omissions, or errors, or items that can be improved. You should tender your criticisms in a constructive fashion: in this manner you will increase the likelihood that people will attend to what you have to say, and your thoughts will perhaps make friends and influence people (rather than the opposite).

On the other hand, there is the occasional reviewer who lets it all hang out. Books seem to have a sort of permanence that papers do not. An incorrect or wrong-headed paper is, after all, ultimately buried in a bound journal volume and hidden away. But a book is always right there on the shelf, staring us all in the face. And, as previously noted, a book reaches out to a larger audience than does a paper. As a result, emotions can run high over a book. I have seen a book review that (literally) began by questioning the editorial decision to publish the book and asserted further that the book completely misrepresented the subject matter; the reviewer spent the rest of the review describing what the subject was *really* (in his opinion) about, with nothing further said about the book itself. And I have also seen a book review [Blo] that compared the subject matter of the book to rather delicate portions of the female anatomy. A recent (and rather controversial) book review [Kli] asserts that the book under review is obviously about a weak subject, as one can see by examining the Bibliography and noting the substandard journals in which the relevant papers are published; the reviewer neglects to point out that he himself is or has been on the editorial board of most of the relevant journals (see [NoS] for an incisive reply). While these essays are briefly diverting they are, in retrospect, embarrassing for us all. As you write your review, pretend that you are reviewing the book of a friend: you want to be honest, and you want to be helpful, but you also want to be scholarly and dignified. Brutality is almost never the order of the day.

In 1978 there appeared a marvelous book on algebraic geometry that is almost universally admired, but that is famous for having a large number of errors: either slight misstatements, or omissions, or incomplete proofs. The fact remains that everyone loves this book, and there is no other like it. (Heck, I may as well tell you: it is [GH].) One reviewer [Lip] praised the book to the heavens, but felt that he had to say something about the hasty writing and the density of errors. So he wrote in part

> If it makes you feel better, think of this book as a set of lecture notes, or even as a fantastic collection of exercises, with copious hints.

Thus the reviewer did his duty: he certainly said something critical, but he said it in a charitable manner, and with good humor. Even the authors must have chuckled over these remarks, and everyone learned something.

The harshest book review that I have ever seen appears in [Mor]. Mordell in fact uses this review to trot out his frustration with the French school's abstraction of his beloved number theory. He attacks not just the book, but he attacks its author on a rather personal and visceral level. A now famous letter was subsequently written by C. L. Siegel to Mordell, praising the review and heaping even more calumny upon the book's author. A discussion of that interchange, and its significance, appears in [Lan]. The trouble with such a review is that any flow of scholarly thought or criticism is lost in the morass of venom and vituperation. No constructive purpose is served by such a review. It is also virtually impossible to have any useful dialogue following upon such a review.

If you are called upon to review a book, and you are tempted to trash it, then I suggest that you set the draft of your review aside for a month (a year would be too long!). Let the ideas gel, and let the words mellow. Show it to a few trusted friends. After a month, you will probably be inclined to take the long view, and to express your ideas in a more temperate fashion. The result will be a better review, and one that you will still be proud of ten years after it appears.

4.3   The Referee's Report

When you are asked to write a referee's report, then you are being requested to offer your opinion as an expert. If you agree to write the report (and you *should*—refereeing is an important part of your professional duties), then you should adhere to the following precepts:

- Write the report in a timely manner—if possible within the time frame suggested by the editor.

- Tell it as you see it. Just as in a letter of recommendation, enunciate your opinion clearly and succinctly, defend it, and then summarize your findings.

- Defend your opinion in detail. You need not find a new proof of each lemma, nor read every bibliographic reference. But you must read enough of the paper so that you can comment on it knowledgeably. While you may not have checked every detail in the paper, you should at least be confident of your opinion as to the paper's correctness and importance.

If somebody asked you whether you liked your car, and whether you would recommend that they buy one, you would not (in all likelihood) tell how each bolt was installed in the chassis, nor how the finish was applied to the body. You would instead summarize the overall performance and features of the automobile. Just so, when you evaluate a paper you should address Littlewood's three precepts: **(1)** Is it new? **(2)** Is it correct? **(3)** Is it surprising? You should speak to its contribution to our knowledge, and to the literature.

- Provide constructive criticisms of the writing, or of the paper's organization. You may enumerate spelling and grammatical errors (if you wish to do so). You should certainly point out mathematical errors, or places where the reasoning is unclear. But you should not be captious. (*Exercise:* Look up this word in your Funk & Wagnall's and think about its relevance to the present discussion.)

- Place the paper in context: How does it compare to other recent papers in the field? Where does it fit? Does it represent progress? If you were not the referee, then is it a paper that you would want to read?

Of course your report should be tailored to the journal to which the paper was submitted. For instance, the *Annals of Mathematics* professes to publish papers of great moment, written for the ages. Other journals have the more modest goal of publishing papers that are correct and of some current interest. Still other journals have no standards at all. You must speak to people in their own language—language that they will understand. Likewise, when you evaluate a paper for a journal, base your assessment on *that journal's value system*.

A typical referee's report is anywhere from one to five pages (or, in rare instances, even more). Its most important attribute should be that it makes a specific recommendation. Everything else that you say is for the record: it is important, but it is secondary.

4.4   The Talk

Giving a talk is different from writing. But it is relevant to the writing process. We ordinarily do some writing to prepare for a talk. And what we write will strongly influence the talk itself. So this topic is fair game for the present book.

A talk is more flexible than a paper. In a talk, you may indulge in informalities, whimsicalities, and a little imprecision; it helps the audience a lot to tell of things tried, and things that failed. You may work trivial examples, and use them as a foundation on which to build ideas.

A talk is also less flexible than a paper. Because the audience receives the talk in linear order—it cannot rewind or speed ahead to check on things—it is therefore at your mercy. You are at a great advantage, when preparing a talk, if you are aware of the limitations of the medium. Endeavor to be gentle.

John Wermer [Wer] makes an excellent case that many mathematics talks are not as effective as they might be because the lecturer is speaking to an imaginary audience located inside his head. This audience is one that knows all the necessary motivation, can pick up on fifty new technical definitions quickly and easily, can follow a technical proof (without explanation) in a jiffy, and can fill in the logical gaps and potholes left by the speaker. Of course such an audience is apocryphal, and thus we are often left with a communication gap between speaker and audience. This section will give you some advice on how not to be like Wermer's *idiot-savant*.

What are the ingredients of a good mathematics talk? First, you must know your subject cold. This does not mean simply that you know it well enough to communicate it to another expert like yourself, but rather that you know it well enough to teach it: that you know the background, the biases, the reasons for the questions, the good and the bad attacks on the problems, and the current state of the art. However, just because you know all these things does not mean that you need to say all of them. A good mathematics lecture is an exercise in self-restraint. Never mind impressing the audience with your profound erudition, your spectacular vocabulary, your extensive professional connections, or your readiness to cite last week's hot results. Instead showcase a nugget of knowledge and insight, and shore it up with crisp comments and incisive examples.

If your talk is scheduled for fifty minutes, then the first twenty should be accessible to a graduate student who has passed the quals. My statement is a strong one. Such a student is not expert at anything. He knows the basics of real and complex analysis, algebra, and perhaps a little geometry. This student has (we hope) an open mind and wants to learn. But your talk in those first twenty minutes should presuppose no specialized knowledge beyond what has just been mentioned. This explains why a nice example or two can be so useful. With an example, God is in the details. The playing field is level, and everyone can benefit. The example(s), of course, should lead to some definitions and the formulation of the questions that you wish to treat in the body of the talk.

The next twenty minutes of the talk should be pitched at a mathematically literate person who is not a specialist. By this I mean that, if your talk is about some part of analysis, then the second twenty minutes should be comprehensible *not just to a specialist in another part of analysis*, but to an algebraist. So you can mention more sophisticated ideas—sheaf theory, or elliptic regularity, or wave front sets, or singular integral operators—but you should not beat them to death.

The last ten minutes can be for the experts, for God, and for you (not necessarily in that order). Every speaker should have a chance to strut his stuff, and the end of the talk is when you should do it. Mention some gory details. Make speculations, formulate technical corollaries, sketch the key ideas in the proof. Forget the neophytes and address yourself to the people who might read your papers. In fact if you do not use the last ten minutes of your talk in this way, then you might leave the impression that you are a lightweight, or that you have nothing to say.

Attempt to finish with a bang. Too many math talks begin with "Well, what I want to talk about today is ..." and then a definition goes onto the blackboard. Too many math talks end with "Well, I guess that's all I have to say" or "I see that I'm out of time so that's it" or "I guess I'll stop here; thank you." Surely you can devise a more creative and informative way to conclude your discourse. You would never end a paper in this fashion. Of course when you write a paper you have time to sit and think of a nice turn of phrase for your conclusion. You should do the same when composing a talk: prepare the introductory sentence or two in advance; likewise prepare the concluding sentence or two. You could finish with a few courteous words of thanks for the opportunity to visit your hosts and to enjoy the hospitality of their department; or you could end with a few mathematical sentences—of real substance— that summarize your enthusiasm for your subject matter. But do end by *saying something*.

The preceding discussion may make it seem that giving a good talk is a piece of cake; that it requires only an acclimatization to certain simple proprieties. Not so. Many other parameters figure into the process.

In fact there are many types of talks: the colloquium, the seminar, and the "job" talk (in which you are showcasing yourself before a department that is considering offering you a job) are three of these. The colloquium is supposed to be for the entire department and perhaps for the graduate students as well; the seminar is for a group of specialists, probably your friends; and the job talk is a set piece—something like Kabuki theater—in which you show yourself. An entire separate book could be written on the art of giving talks. In the interest of brevity, my remarks below will center around colloquium talks. Seminar talks are less demanding and job talks more so. The remarks below apply in some form to *any* talk; the trick in interpreting my advice for a particular circumstance is to *know your audience*. As you read my detailed remarks below, keep this unifying principle in mind.

1. Showcase *one theorem*, or perhaps a single cluster of theorems. There is no point to giving a talk on five truly different theorems, because the audience cannot absorb so much material in one sitting. On the other hand, if you cannot build your talk around one theorem then perhaps you have nothing to say. Here is what Gian-Carlo Rota has said about the matter:

Every lecture should make only one main point. The German philosopher G. W. F. Hegel wrote that any philosopher who uses the word "and" too often cannot be a good philosopher. I think he was right, at least insofar as lecturing goes. Every lecture should state one main point and repeat it over and over, like a theme with variations. An audience is like a herd of cows, moving slowly in the direction that they are being beaten into. If we make one point, we have a good chance that our audience will take the beaten direction; if we make several points, then the cows will scatter all over the field. The audience will lose interest and go back to the thoughts they interrupted in order to come to our lecture.

If the talk is a survey, then you should temper this last advice to suit the occasion. Better to give a survey of a particular aspect of semi-Riemannian geometry than to endeavor to survey the entire subject of geometry. And do suit the talk to the audience. You can survey non-Euclidean geometry for junior/senior mathematics majors, or you can do it for seasoned mathematicians. But you would do it differently for each of these audiences.

2. Have an attractive title. A casual observer, seeing the title "Subelliptic estimates for a quasi-degenerate, semilinear partial differential operator satisfying a weak symplectic condition with applications to the hodograph technique of Hörmander," will probably be more tempted to head for a late afternoon beer than to attend your talk. The title "A new attack on a class of nonelliptic equations" conveys the same spirit and is likely to suggest to a broader class of people that the talk may contain something for them.

3. *Prove something.* It leaves a bad taste in everyone's mouth if you talk about a subject but do not get in there and do it. One good strategy is to prove a special case, or work out an example, in some detail; then use this prolegomena to sketch the key ideas in the proof of the main result.

4. Structure your talk so that everyone will take something away from it. Ideally, a member of the audience who is questioned that evening about that day's colloquium should be able to say "The talk was about this" or "The main theorem was that" or "The speaker was relating geometry to combinatorial theory in a new way." If you bear this thought in mind while composing your talk, then it will have a strong, and salubrious, influence on your entire approach to the process.

5. *Be specific.* Heed this advice, both when you are writing and when you are speaking. Nobody wants to listen to an hour of vague fluff. Nobody wants to perceive that you are dodging the main point of the discussion. If you appear to be evasive then, at best, you could make people think that you are sloppy and imprecise; at worst, you could leave people with the impression that you are faking it—indeed that you do not know how to prove these theorems.

   I once heard a mathematician dedicating a lecture to an eminent person, on the occasion of that man's sixtieth birthday. In brief, the dedicator said "In my country the tradition in lectures has been to deal in vague generalities. This man has taught us to present examples, and to work through them completely." The value of showing your audience the inner workings of the material you are presenting cannot be over emphasized. This process helps to draw in students (both young and old), and shows them how the subject works. It also helps to involve those who are not already expert.

6. Do not be afraid to dream. I say this cautiously, for I have already warned you not to prevaricate or mislead. But a talk is a different vehicle from a formal piece of writing. Standing before a group and speaking is an opportunity for you to tell the audience what you tried, what did not work, and what might work in the future. It is absolutely impossible in mathematics to publish a paper that says "Today I woke up and tried to prove the Riemann hypothesis and I failed." In a mathematical *talk*, you can

dandle such thoughts before your audience and not only survive, but in fact heighten the audience's appreciation for you and your insights.

7. Do not be afraid to be informal. One of the most effective devices that I have seen is for the speaker to say "If we assume these three explicitly stated hypotheses, together with some other technical things that I shall not enunciate, then the following conclusion holds." Often the technical items that are left unspoken are of great interest to the deep-down experts; but to everyone else they would be meaningless, indeed confusing. It takes real insight, and a dash of courage, to be able to say to the audience that you are sloughing over some difficult points. Of course you should never lie; but you may certainly downplay some of the technical points in your subject.

   These comments also apply when you are presenting a proof. In a specialized seminar, it might be appropriate to slug your way through every technical detail of your argument. In a colloquium, such arcana are virtually never appropriate. If your theorem has any substance at all, then its proof may consist of ten or twenty or more pages of dense argument. It could take a serious reader a week to digest thoroughly the inner workings of your reasoning. Thus it could never work to present the entire theorem, with its proof, in a colloquium talk. Hit the high points, say a word about what you are omitting and over simplifying. Proud as you are of the cute argument you cooked up for the proof of Sublemma 3.1.5, do not trot it out during your colloquium.

8. Prepare your talk with multiple entry points and multiple exit points. What does this mean? Rare is the listener who can pay rapt attention for the full space of 50–60 minutes. Many members of your audience will drift in and out. If you say something interesting, then certain people will begin to think their own thoughts, or try to produce a necessary example or lemma. Make it easy for such people to "re-enter" the lecture. Provide several doorways.

Likewise, there is no way to predict how a given talk will go before a given audience. If you are lucky, there will be fortuitous interruptions and serendipitous comments. Time will not be used in just the way that you had planned; you could easily be caught short. Therefore you should create several propitious points at which you can make a gracious exit from the talk. As already noted, a hasty "Egad, I'm out of time" is not a savvy way to finish your colloquium. In any event, do not run overtime—at least not by more than a couple of minutes. First, to do so is rude; second, colloquia are at the end of the day and people have other things to do (such as going home to dinner); third, people simply have no patience for a talk that exceeds the allotted time. At my university, we had a leading job candidate who, in his ceremonial talk, ran out of time. He gave us a big smile, went to the clock, and pushed the big hand back twenty-five minutes. And then he used them! Suffice it to say that there was no further discussion of his candidacy.

9. Prepare, prepare, and prepare some more. You should have thorough notes before you, but you should rarely refer to them. Your talk should have an edge: you want to be thinking through the ideas with your audience, and you want to be *talking* to the people in the room. You are not giving a recitation to your buddy in the front row. You are not lecturing to the fictitious audience that is engraved in your frontal lobes. You are talking to the individuals who are breathing the same air as you. Pick them out as you speak; look at them; change your focus and your depth perception as the talk develops. Pace around. Step backward and forward. Use your body and your voice to lure the audience into the talk. Do *not* be a slave to your notes.

Several years ago I watched an eminent mathematician prepare to give a colloquium on a topic that I personally had seen him lecture on at least four times previously. He had probably lectured on it fifty times in total. I had attended his course on the subject. He *owned* this material; he had created it from whole cloth. He could have given this talk in his sleep. Nonetheless for this, his fifty-first performance, he insisted on sitting quietly in a room for an hour and writing out everything that he was going to say. During the talk, he cast not a glance at his notes. At the end of the talk, he summarily dumped the notes into the trash.

This process made a tremendous impression on me, and I have reminisced frequently about what I observed. *Writing out his talk was his mantra.* He used this process to prepare himself psychologically for the talk. Some people will prepare by just staring off into space and walking, mentally, through the talk. Others will stroll to the student center and buy a cup of coffee. Still others will spend the entire afternoon in the library sweating over the literature and worrying about questions that someone might ask but in fact never will. It does not matter what you do to psyche yourself up for your talk, to guarantee that you are prepared. The main point is to *do something*: find a technique that works for you and use it.

10. Be careful in your talk to give credit where it is due. Do not give attributions only when your name is involved. In fact most speakers tend to do the opposite. When presenting someone else's theorem, the speaker is careful to write out all the relevant (sur)names in full. When it comes to his own theorem, the speaker just writes something like

> **Theorem:** [Fu-Isaev-K]
>
> Let $\Omega \subseteq \mathbb{C}^n$ be a smoothly bounded, pseudoconvex domain with noncompact automorphism group ...

This citation is an example taken from my own life, where Siqi Fu and Alexander Isaev are my collaborators and "K" is yours truly.

Now that I have listed the ten commandments, let me turn to a discussion of general principles. Many technical skills are necessary for giving a good talk. I have already mentioned eye contact and organization. Let me also discuss blackboard technique. Even if you are a great expert in your subject, and have a charming and erudite delivery, you will be putting a substantial barrier between yourself and your audience if your writing is an incoherent mess, or if you fill the blackboard with a chaotic barrage of longhand. Learn to write in straight lines, horizontally, from left to right. Write large, and write neatly. Do not put much on each blackboard. Give the audience a chance to copy what you have written before you erase it.

Do not stand in front of what you have written. As you write, read the sentences aloud. Learn to draw your figures accurately and skillfully. Isolate material that will require later reference and *do not erase it.* Plan in advance how you will use the blackboard, so that you can be sure that you will always have room for what you want to write next. Just as the director of a play knows in advance where each actor will be at each moment, and plans every movement on the stage in considerable detail, so you should plan the moves of your talk in advance. The audience will grow phenomenally frustrated watching a forlorn speaker pace back and forth in front of his blackboards—for several minutes!—trying to decide what to erase, or what to save.

Some people solve the blackboard problem by using overhead slides (transparencies) instead. The very act of creating slides in advance addresses virtually all the issues that I have raised about blackboards in the last paragraphs. Slides, in the hands of a skilled user, can be a powerful tool. (The blackboard is sometimes inescapable, however, so you should learn to use it.)

If you do use slides, then learn to use them wisely. Each slide should contain one thought, or one idea. Each slide should contain about six to eight lines, and should have wide margins. The bottom 2 inches of each slide should be left blank—because this portion of the slide is often blocked from the vision of those in the back of the room.

Do not write out complete, long sentences on your slides. Abbreviate wherever possible. If you are going to TeX your slides, then consider using SliTeX (which is a version of TeX that contains extra large fonts and other artifacts that are useful for preparing overhead slides). Note, however, that a neatly prepared handwritten slide is often as effective as a TeXed slide—and handwritten slides give you the additional flexibility of colors, arrows, and other graphic tricks. You should have only about one slide per two to three minutes of speaking.

I have seen talks in which the speaker simply printed out the text of a fifty-page TeX document onto transparencies—in 10-point or 12-point type. Moreover, the speaker showed every single slide to the audience. What a disaster! First, this is far too much material per slide—and none of it can be read. Second, this is too many slides for a fifty- or a sixty-minute talk.

One of the most important skills that you need to develop, both as a teacher and as a colloquium or seminar speaker, is time management. You need to fit what you have to say into the time allotted. People will be monumentally irritated to watch a mature mathematician spend the last thirty minutes of his fifty-minute talk pacing back and forth, scowling at the clock, and declaiming that he does not have sufficient time to present his thoughts. I have seen many such a speaker act as though it were the audience's fault, or the university's fault, or perhaps his host's fault, that he did not have more time. What nonsense. The speaker knew when he was invited—probably many months before—what the parameters were. Giving a fifty- or a fifty-five-minute colloquium talk is part of the academic game. Learn to play by the rules.

Perhaps you are saying to yourself—or have said to yourself in the past—"all good and well, but this speech-making stuff is for joke-tellers and hams and showoffs. I am a scholar. I am not an actor." Such a statement is a cop-out (if you will pardon the vernacular). Nobody expects you to be Bob Hope. Part of a scholar's existence is to communicate—both in writing and in speaking. The thoughts in this section are intended to help you to enhance your abilities with the latter. Giving a talk is a personal affair; you should do it in the fashion that best suits you. But I hope that the ideas presented here will help you to sharpen your wits and your technique.

4.5 Your Vita, Your Grant, Your Job, Your Life

The Curriculum Vitae

A businessman has his resumé and an academic has his Curriculum Vitae (or *Vita* for short). The Vita is your professional history—it should give a quick sketch of who you are, where you were educated, your professional experience, any honors that you have earned, your scholarly accomplishments, and related materials. Usually you will include your publication list with your Curriculum Vitae.

Your Vita should *not* read like this:

> Born on a mountain top in Tennessee.
> Greenest state in the land of the free.
> Raised in the woods so he knew every tree.
> Killed him a b'ar when he was only three.[2]

All quite charming, but a Vita should *never* be in paragraph (or stanza) form. The material should be laid out in a tableau so that the reader can quickly pick out the information he needs. Your name should be in boldface at the center top. (I recommend that you use your official name—the one on your birth certificate. Your friends may call you "Goober", but you should save that information for another occasion.) Quickly following should be your date of birth, your educational information, your address and phone numbers and *e*-mail address, your employment record, key honors earned, and so on. An example of the first page of a Vita appears later in this section.

Your publication list should be a separate section of the Vita. Those who are especially careful separate published works from unpublished (or to-be-published) works and separate items in refereed journals from items in nonrefereed publications (such as conference proceedings); books are often listed separately; some people even list class notes they have prepared, or software that they have written (if you are a numerical analyst or a specialist in algorithms then this last would be essential). Usually the items in a publication list are given in approximate chronological order, although some people use reverse chronological order.

Another section of the Vita lists grants or funding that the person has received over the years. For each grant, you usually list the funding agency, the title and number of the grant, the amount of money in the grant, and the year(s).

Often a Vita will include a section of invited talks or, if you are quite senior, of major invited talks (that is, an hour speaker at a national

---

[2]From *The Ballad of Davey Crockett,* Walt Disney Productions.

AMS meeting, or principal speaker at a CBMS conference, or a speaker at the International Congress).

Yet another section will list graduate students (Masters and Ph.D.) that you have directed. Another could list material describing your teaching experience (courses taught, curricula developed, and so forth). Indicate your expertise with computers—either software developed, or courses taught, or other accomplishments.

Certainly say something about your skill with foreign languages. Have you done any translation work? Are you well traveled? Have you taught in another country?

Finally, some Vitae have a catch-all section with editorial activities, service to professional societies, or anything else that the person writing the Vita thinks may be of interest.

Your Vita is no place to be humble. This document is the *gestalt* that you present to the world. Certainly do not prevaricate—or even exaggerate—but be sure to tell the reader everything that you want him to know about yourself.

At the risk of sounding preachy, let me expand a bit on one of the points in the last paragraph. When preparing the Vita, we all want to present ourselves in the best possible light. There is a tendency to dress things up—beyond what is strictly kosher. Perhaps you did not complete that French course—but you ate quiche Lorraine once— so you write that you speak French. The letter from the journal to which you submitted your latest paper says "if you make the following twelve changes then the referee will have another look at it," and you list the paper as accepted. The NSF tells you that you are on the "maybe" list for a grant, and you put on the Vita that you have a grant. People who have made these slips are not liars; they are just trying too hard. Strive to avoid such exaggerations. Most departments check facts carefully. Many schools only believe in publications that have appeared, and for which there is a *bona fide* reprint (many schools have been burned once too often in the past). If the Funded Projects office at your school does not have the letter from the NSF, then your grant does not exist. Worse, if you make such claims in your Vita and the claims do not wash with the people evaluating your case, then the situation will weigh against you. My advice is to be extra careful.

`SAMPLE VITA`

CURRICULUM VITAE
for
Clemson Ataturk Kadiddlehopper

Date of Birth: March 15, 1947

Home Address: 17 Poverty Row, Faculty Ghetto, Iowa 50011

Current Academic Affiliation: Department of Mathematics
Walmart A&M, Sam's Clubville, Iowa 50012

Telephone:	(515) 294-6021 (office)
	(515) 373-3286 (home)
	(515) 294-6047 (fax)
e-mail address:	`CAK@MATH.SAM.EDU`
Graduate Education:	*Ph.D., Mathematics*
	Montana Institute for the Tall, 1974
	Thesis directed by Charles Ulmont Farley
	M.S., Mathematics
	Frisbee State University, 1971
Undergraduate Work:	*B.A., Mathematics*
	Joe's Bar and University, 1969
Academic Positions:	Assistant Professor, College of the Yodeling Yuppie, 1974–1979
	Associate Professor, Steland Lanford University, 1979–1988
	Professor, Walmart A&M, 1988–present

Honors: Neural Sediment Fibration Graduate Fellow, 1971–1974
Visiting Professor, Callipygean Institute of Tectonics, 1977
Shinola Fellow, College of Good Hair, 1979
Visiting Professor, Upper College of
      Lower Academics, 1980
Visiting Professor, University of Basic Bourgeoisie, 1986
Visiting Professor, Hahvahd University, 1986
Honorary Lecturer, Crab Louie College, 1987

Now let us return to matters prosaic. You must tailor your Vita to the circumstances. I have been teaching for 25 years. Thus it would be crazy for me to list every course that I have ever taught. It would make more sense for my Vita to list courses that I have created, or textbooks that I have written. On the other hand, if you are just starting out in the profession, then you should indicate the depth and range of your teaching experience and certainly indicate your facility with computers, both in the classroom and outside it. A beginner will probably have directed few if any graduate students. That is not a problem, since such activity is not expected. Do, however, be complete in describing your other activities.

The Grant

Funding is available for many different types of activities that a mathematician might undertake. These range from quite specific, goal-oriented projects that are funded by industry all the way to grants from the NSF (National Science Foundation) to encourage pure research in abstract mathematics. There is also funding from the Department of Defense, from DARPA (an arm of the CIA or Central Intelligence Agency), from NASA (the National Aeronautics and Space Administration), from NIH (the National Institute of Health), from DOE (the Department of Energy), from NSA (National Security Agency), and from many other sources as well. Granting agencies such as the NSF have considerable funds to encourage work on the mathematics curriculum—from developing new ways to teach calculus to revising substantial blocks of undergraduate mathematics education.

Given the range of activities that granting agencies are willing to fund, and given the variety of different potential sources of grants, I could discuss grantsmanship at length. I shall content myself here with a few general precepts that should apply to virtually any grant application that you may write.

Read the prospectus for the program to which you are applying. Doing so, you will learn what the program is looking for, what particulars should be itemized in the proposal, what page limits will be enforced, and when the proposal is due. Learn about what type fonts are acceptable, what margins the pages should have, how long the Curriculum Vitae portion of the proposal should be, how long the references section should be, how many pages should address previous work, how many pages should address new work ... *and so forth.* Grant proposal writing is not a free form activity. Get the rules straight before you begin.

The main issue that is in the air when your grant proposal is being evaluated is your credibility: *can* you do the work that is being proposed, and *will* you do the work that is being proposed? Given your stature, your abilities, and your track record, is it clear that you can work on these problems (be they research or education)? Can you solve them or make progress on them? Are you capable of evaluating your own progress? Finally, can a case be made that you are *the right person* to work on this project? Or will the work be done as a matter of course by others (if indeed it is worth doing at all)?

You must walk a delicate line here. On the one hand, you want to make it clear to the potential granting agency that you know this subject inside and out, that you know the existing literature, and that you have a good program for proceeding. You want to demonstrate that you are already engaged in some version of the proposed activities. On the other hand, you do not want to make it sound as though you have already solved the proposed problem. You also want to give the strong impression that you are working on substantial problems of real significance; these should be problems for which even partial results will be of interest. But it should be plausible that you are up to the task. In particular, if you propose to prove the Riemann hypothesis, then you will have a difficult time making your case. After all, many of the bigshots are working on this problem; if they cannot make inroads then how will you?

Generally speaking, granting agencies will not provide funds to help you to learn something new, or to retool. Thus you should make a case that you are already engaged in the proposed project, that you have a grip on it, and that you have a viable program. If those considerations entail your learning something about nonlinear elliptic PDE's, then by all means you should say so. But a grant proposal that reads (in effect) "I'm tired of studying finite groups so now I'm going to do symplectic geometry" just will not fly.

As already noted, you must prove that you are up on the relevant field—not just what is in the published literature but what is available in preprint and other tentative form. For the most part, grants are refereed by your peers. These will be peers who are on the cutting edge. They will judge you by their own standard—the standard by which they themselves would expect to be judged.

When writing a grant proposal, you must walk another delicate line. You will usually have a page limit. You simply cannot go on at length, or have extensive digressions, or have verbose introductions or chatty conclusions. But you must make the proposal as easy to read, and as self-contained, as possible. If your proposal engages the referee's interest, and teaches him something, and does not force him to keep running to the library to figure out what you are talking about, then you will be at a real advantage. If, instead, your prose is a bore and the referee has to slug his way through it, then your proposal is likely to be penalized.

Do not be afraid to telephone the granting agency to which you are applying and to talk to the program officer. Many grant programs, and many program officers, encourage this activity. By talking to a program officer, you can better focus and tailor your proposal to the goals of the intended program, and you will not waste the program's time with a proposal that is completely off the wall.

Proposals in mathematics education and curriculum often require a section on "self-evaluation" and a section on "dissemination". You should not (in the self-evaluation portion of your proposal) say "We'll see how happy the students are at the end by distributing teaching evaluations." You also should not (in the dissemination portion of your proposal) say "I'll go to conferences and talk about this stuff with my buddies." I have seen both of these in serious proposals, and they do not work. Both approaches are too facile, and show no imagination and no effort. Good self-evaluation programs often involve motivational psychology experts from your institution's School of Education, tracking of students after they leave the experimental program, exit interviews, and many other devices. Good dissemination programs often involve writing a textbook for publication, creating a newsletter, setting up a web page, organizing workshops, and so forth. I am not necessarily advocating any of these devices. I am merely explaining how the world works.

Self-evaluation and dissemination play an implicit role in a research proposal. Your report on previous work will give an indication of your ability to evaluate your own progress. The scientist who says "In the last five years I tried a lot of things but nothing panned out" shows both bad judgment and an inability to learn from his work. The dissemination aspect of a research proposal is reflected in your publication record, your invitations to speak at conferences or colloquia, and your collaborative activities.

Before you submit your proposal, run it through a spell-checker. Check and recheck the grammar. Show it to a senior colleague. Proofread it more times than you think could possibly be necessary, and then proofread it once more. The reviewer will be phenomenally irritated to read a proposal that appears to have been prepared hastily or sloppily. Your proposal should be as slick as glass. It should be a pleasure to read, and it should get the reviewer excited about and interested in what you are proposing to do.

Your Job

At one time or another, most of us will have to apply for a job. Let me first say a few words about applying for an academic job.

When you apply for a job at a college or university, you send in your Vita (discussed above) and a cover letter. The cover letter should be brief (well under a page); it should identify you, your present position, the type of position you seek, and your areas of interest. No more. A cover letter with a multitude of exclamation points, shamelessly extolling your virtues as a teacher and your bonhomie as a colleague, is highly inappropriate. A sample cover letter appears later in the section.

If you are a beginner in the profession, with a short publication record, then you might include some of your preprints with the job application. Your Vita should also list your "references" or "recommenders". This point is vital, and many job applicants overlook it. Before applying for any job you should approach three or four people (for a senior job it could be six or eight, or even more) and ask whether they are willing to write in support of your application. Ideally, these should be prominent people in your field, whose names will be recognized by those evaluating your dossier. Once you induce a group of such people to agree to write,[3] then include their names, business addresses, business phone numbers, e-mail addresses, and fax numbers in a section of your Vita called "References". These days—especially if you are a job candidate just a few years past the Ph.D.—you should have one or two letters from people who can say something specific and positive about your teaching. Their names should be listed in your References section as well. It is now commonplace for job candidates in the United States to include the "AMS Standard Cover Sheet" in the dossier; this form may be obtained from most issues of the *Notices* of the AMS.[4]

Some job candidates arrange to have a sample of their teaching evaluations, or passages from their present institution's Teacher Assessment Book (such a book is often published by the campus student organization) to be included in their dossier. Such an inclusion can help to lift the dossier out of the ordinary, and will add substance to the letters that praise the candidate's teaching abilities.

If you use your imagination, you can probably think of all sorts of things that might be included in your dossier in candidacy for a job. I recommend that you consider each one cautiously. People on hiring committees these days often must wrestle with 500–1000 job applications in a season. A big, fat dossier will just turn off a weary committee member. So do not leave anything important out of your dossier, but think carefully about what you do include.

If you make your application in the manner described in the preceding paragraphs, then a school to which you apply will know just how

---

[3]The notion of the candidate *asking* people to write for him seems to be a peculiarly American custom. In many countries—especially in Europe—the hiring institution does all the solicitation of letters.

[4]Many people, especially young job candidates, include in their job application a one or two page statement describing their research program; many young people also include a "statement of teaching philosophy." The first of these can be quite helpful to a nonexpert who is endeavoring to evaluate the dossier: a good research statement can at least guide such a reviewer to an appropriate expert colleague who can comment on the case in detail. I dare not say whether a "teaching philosophy" statement has any real value; there is little grass growing in this subject area, and you will have trouble finding anything interesting or original to say. On the other hand, some schools *require* a statement of teaching philosophy; in such a circumstance you must do your best to write something thoughtful and thought provoking.

to process your paperwork. Once it has your cover letter and Vita, it can start a file on you. Then it has a place to put the letters of recommendation as they come in. And, because you have included a list of references, the school will know when your dossier is complete.

If your application is for any position beyond a beginning lectureship, and if you make the "short list", then you will likely be invited to give a talk and to meet your potential future colleagues. Let me not mince words: this is a make-or-break situation. Dress well (not as though you were entering a ballroom-dancing contest, but rather as though you are taking the situation seriously). Give a polished, well prepared talk (see Section 4.4 on how to give a talk). Think in advance about some of the topics of conversation that may come up when you meet your new colleagues. Be prepared to describe your research conversationally to a small group of nonexperts; be able to say in five or ten minutes what you do, how it fits into the firmament, who are some of the experts, what are some of the big questions.

Be prepared to say who in this new department has interests in common with yourself; with whom you think you might talk mathematics; who might become your collaborator. Do not underestimate the significance of this circle of questions. Do *not* say "Oh, I talk to everybody; I'm the Leonardo da Vinci of modern mathematics." Such a statement is not credible; utter it and you will surely send yourself plummeting to the cellar of the short list.

Think over your ideas about teaching, about the teaching reform movement, about teaching with calculators or computers, about teaching students in interactive groups, and about any other topics that may arise. Some schools have special problems connected with the teaching of large lectures; be prepared to share your views on that topic. Other schools have special tutorials for calculus students; be prepared to chat about that topic as well. Some schools like to conduct a formal interview, with a few of the senior faculty asking you direct questions about your research, your teaching, your attitudes about curriculum and reform, about teacher/student rapport, or anything else that is in the air at the time. It makes a dreadful impression if you are inarticulate, do not seem to know your own mind, or simply have not given any thought to these matters. I am not advocating that you go to your ceremonial

SAMPLE COVER LETTER

November 22, 1996

Noodles Romanoff, Chairman
Department of Mathematics
Little Sisters of the Swamp College
Sanctuary, Oklahoma 23094

Dear Professor Romanoff:

I wish to apply for a faculty position, at or near tenure, in your department. I received the Ph.D. in Mathematics in 1974. I am a geometric analyst, with specializations in complex function theory, harmonic analysis, and partial differential equations. My Vita is enclosed. It includes my list of references. I currently hold the position of Instructor of Mathematics at Brouhaha Subnormal School in Wichita Falls.

Please note that my research is supported by a grant from the Normative Sodality Agency. I am also a recipient of the Mudville Distinguished Teaching Award. I have strengths in research, teaching, and curriculum.

I look forward to hearing from you.

Sincerely,

Shrimp Chop Suey
Instructor of Mathematics

job interview with a sheaf of notes *in your hand;* I am instead advocating that you go with a few note cards *in your head.*

When a school decides to offer you a job, the chairman will usually telephone you, or send you an *e*-mail message followed by a phone call. At that time he may discuss salary, teaching load, computer equipment, health insurance, the retirement annuity, or other perquisites. At that time, you may wish to ask about these or about other concerns.

Important information about you can be lost in the Vita—especially if your Vita is long. If you are a graduate student applying for a first job, and if you have won a teaching award for "Best TA", then certainly mention that encomium in your cover letter. If you are a few years from the Ph.D. and the holder of a Sloan Fellowship, or an NSF Postdoc, then you should mention these honors in your cover letter. You do not want your cover letter to look like a flyer for your local supermarket, but you want it quickly to lead the reader to your strong points.

If you are applying for a job in the private sector—say at Texas Instruments, or AT&T, or Aerospace Corporation—then the application process may be a bit different from the process in an academic setting. An industrial organization is probably not interested in letters of recommendation from the Université de Paris, nor in binary comparisons with famous young algebraic geometers. A business "resumé" is different from an academic Vita. Go to your local bookstore and purchase a book on how to write an effective resumé (see, for example, [Ad3]), how to write a cover letter (see [Ad1]), and how to apply for a job (see [Ad2]). Although working in industry certainly will involve communication skills, it probably will not involve much classroom teaching. The interview for an industrial job will likely be even more crucial than the interview for an academic position. Consult acquaintances who have been through the process so that you can be well prepared.

Your Life

In the abstract, the rewards for good writing may seem far off and vague; instead you can see clearly how a well-written Vita or grant proposal could lead to just deserts. Your Vita is a tool for helping you to find employment, or a promotion, or to achieve some other goal. Your grant proposal is a way to seek funding. I have also discussed how to find a job. The principles of good writing described in other parts of this book apply just as decisively to these practical matters: express yourself directly, cogently, and briefly; do not show off; know what you are talking about; and (paraphrasing Jimmy Cagney) plant both feet on the ground and tell the truth.

I have seen many a Vita in my time. One of these contained a page entitled "Cities Beginning with the Letter 'Q' in which I Have Spoken Fewer than Five Times." Another listed an uncompleted mystery novel. Yet another listed forty (count 'em) collaborative papers that were incomplete and in progress. One Vita by Mathematician X listed poetry, both published and unpublished, that was written to X, by X, about X. Another Vita listed the subject's (not very happy) marital history. Your Vita is a business document. This piece of paper is a précis of your professional life. Think carefully about what you put into it and how you organize it.

Your grant proposal is a manifestation of your professional values, what you are all about, and what you are trying to do. As you develop it, read it with the eyes of your potentially most critical reviewer.

Here we are discussing writing with immediate impact, and with a direct effect on your life. This is writing that you wish to succeed because it must. Even more than in your other writing, you will want to strive to make each word count, and to force each sentence to say precisely what is intended. The critical skills discussed in this book should help you in these tasks.

4.6 Electronic Mail

For many of us, electronic mail (or *e*-mail for short) has become an important part of life. The technology of *e*-mail has enabled us to carry on extended conversations with people all over the globe. We can engage in topic-specific discussion groups, conduct business, develop friendships, and even have fights via the Internet. Perhaps more significantly, we can conduct mathematical collaborations with people 10,000 miles away, in some cases with people whom we have never even met. You may actually (though I encourage you to exercise this option with discretion) send an *e*-mail blind to a professor at MIT and say "Hello,

I'm so and so. Do you know the answer to the following question?" I have occasionally engaged in this speculative activity and, more often than not, I have received a useful answer.

I am currently writing a series of papers with two collaborators, one of whom is usually in Los Angeles and the other usually in Canberra, Australia. During this last year, one of us spent a leave in Berkeley, another took a leave in Wuppertal, Germany, and the third changed jobs. We did not miss a beat, because e-mail is oblivious to these moves. Marshall McLuhan [McL2] died too soon: the global village is finally here in spades.

G. H. Hardy and J. E. Littlewood carried on what is by now the most famous, and certainly the most prolific, mathematical collaboration in history. Usually in two different locales (one in Cambridge, the other Oxford), they conducted their collaboration by regular post (now known as "snail mail"). Their hard and fast rule was that if one of them received a letter from the other, he was under no obligation to open it—right away or at any time. Many a letter was thrown into a pile, not to be read then or perhaps ever; this to guarantee that the recipient could think his own thoughts, and not be interrupted. To my mind, e-mail is a bit different: it gets right in your face, once or several times a day. Once you have determined what a particular e-mail message is and whence it came, then you are looking at it. The temptation is to read it. As McLuhan taught us [McL1], "the medium is the massage."

For many purposes, communicating via e-mail is preferable to communicating by telephone. For an e-mail message has the immediacy of a telephone call without any of the hassle of playing "telephone tag", talking to voice mail, or patiently explaining your quest to a secretary. Many of us find that we send more e-mail than we do letters, and we use e-mail more often than we use the telephone.

Because the use of e-mail has become so prevalent, we must all learn some basic etiquette of the e-mail system. As with many other activities in life, e-mail is something that we can benefit from if we give it just a few moments of reflection.

- Be sure that your e-mail messages go out with a complete header. This header should include a "From" line, indicating your identity and *e*-mail address, and a "To" line, indicating the identity and *e*-mail address of the person to whom the message is being sent. It is not mandatory, but is highly desirable, for you to include a "Subject Line" in the header. Many busy people receive 50 or more *e*-mail messages per day; you do such people a great favor by helping them quickly to identify and sort their *e*-mail.

Some systems allow you to strip away all or part of the header of your *e*-mail message. I urge you not to do this as such an action is unprofessional and rude. Sending anonymous *e*-mail is no better than sending anonymous hate mail.

- I often receive *e*-mail messages that say (*in toto*) "Yes, I agree with you completely" or "Right on" or "There you go again!" I love fan mail as well as the next person, but I often cannot tell what such *e*-mail is about. Do yourself and your correspondent a favor and either **(i)** include the *e*-mail message to which you are responding in your reply or **(ii)** at least include a sentence or two indicating to what you are responding.

- Sign your *e*-mail message with your full name. Signing off with "See you later, alligator" or "That's all, Folks" is momentarily amusing, but it often forces your recipient to search the header of the message to determine whose pear-shaped tones he is reading. Such a search is sometimes frustrating, and irksome to boot.

The best possible "signature" to an *e*-mail message is something like this:

```
******************************************************
* Steven G. Krantz  (314) 935-6712  fax (314) 935-6839*
* Department of Mathematics, Campus Box 1146         *
* Washington University in St. Louis                 *
* St. Louis, Missouri 63130-4899  SK@MATH.WUSTL.EDU  *
******************************************************
```

Of course you do not want to type out this mess each and every time you send an *e*-mail message. The operating system UNIX makes it easy for you to avoid such tedium. Here is how to do it if you are using the mail utility PINE (if, instead, you are using the built-in UNIX mail utility, then the steps are similar but not identical; in ELM you need to learn appropriate commands for the text editor that you are using):

(1) First, create a file containing (your version of) the box of text indicated above; let us say that the file is called SIGN;

(2) Store the file SIGN in your home directory. Next time you send an *e*-mail message then, after you finish typing the message (but before you send it!), perform the following sequence of steps:

(a) Type <Ctrl>-r

(b) Type SIGN

(c) Press <Enter>

When you complete the third step, the file SIGN will be read in at the cursor position in your *e*-mail text. Now you can send the message.

This sequence of steps will become second nature for you after a few tries, and I encourage you to develop the habit. Your correspondents will also appreciate your consideration. We will also have occasion to refer to this protocol when we discuss including a TeX file or other file in an *e*-mail message.

On many systems there is a useful shortcut that you can implement with PINE and other *e*-mail utilities. Use your favorite text editor to create a file called .signature . Put your "signature" text in this file. Place the file .signature in your home directory. Now, whenever you begin an *e*-mail message, this signature will automatically be appended to it.

Note that certain USENET newsgroups and other Internet nodes ask you to restrict your SIGN or .signature file to four lines. If you use *Eudora*, or one of the many third-party Internet providers, then the "Signatures" menu performs the above operations for you automatically.

- The good news about *e*-mail is that it is a lot like conversation. It is spontaneous, natural, and candid. The bad news about *e*-mail is that it is a lot like conversation—without the give and take of an interlocutor. Thus we are tempted to type away madly, at high speed, having no care for corrections or proofreading. This is a big mistake.

Proofread each *e*-mail message before it goes out. If the message is important then proofread it several times. Most *e*-mail editors are easy to use. In the UNIX environment you have a choice: the PINE editor is self-explanatory, and much like a word processor; in the ELM environment you can customize the editing environment, using EMACS, or VI, or another editor of your choosing. In any event, learn to use the editor on your system and *use it.* Correct misspellings (many an *e*-mail editor is equipped with a spell-checker) and misstatements. Clean up your English. Some *e*-mail messages that you send will have the permanence of a hard copy written letter. Send something that will reflect well on you.

In fact, when I am writing something of great importance, I compose it on my home computer—on the text editor with which I am most familiar (see Section 6.3 for a discussion of text editors). I do this in part for psychological reasons. When I compose on my home computer, I do not worry about the system hanging or going down; I do not worry about taking a break and being thrown off the system; and I am using a writing environment with which I am thoroughly conversant. I can use my spell-checker, my CD-ROM dictionary and thesaurus, and other familiar resources to put the document in precisely the form that I wish. I also can sleep on the matter before I send the document.

The next morning, I bring the document to work on a diskette, upload it to the system (see Section 6.9), and then pull it into an *e*-mail message using the protocol described above for the file SIGN. This methodology is a valuable tool.

- Implicit in the preceding discussion is a major liability of *e*-mail. Too easily can you write something in haste in the *e*-mail environment and then just send it off—it only requires a key stroke or two!—and then it is gone. You cannot retrieve it.

  I once had a rather significant fight with another mathematician. He wrote me a letter taking me to task for something that I had done. Fortunately, this event occurred in the days before *e*-mail. I wrote a hasty and heated response (in hard copy, for that was all that we had at the time), and dropped it in the department's outgoing mail tray. An hour or two later, I pulled the letter from the mail (I had been stewing about it all the while), and penned a milder version of the heated letter. This revision process repeated itself throughout the day. By the end of the day, I had put in the mail a letter of apology, acknowledging my error and thanking my correspondent for calling it to my attention. I have always been happy for this outcome. With *e*-mail the outcome would have been quite different.

- Try to keep your *e*-mail messages brief. Of course I realize there are times when you are circulating a report or writing a detailed formal analysis of some situation; in such circumstances, it may be appropriate to go on at some length. But, most of the time, when writing *e*-mail, you are sending a memo. Thus make it quick. Often, on the computer, we tend to do things just because we can. Writing an *e*-mail message is a lot like talking, but without the reality check of having someone interrupt you from time to time. Thus you must show some good sense: say what you have to say, say it cogently and completely and *concisely*, and then cease.

- You can easily forward any *e*-mail message that you receive to anyone that you like. I am astonished at the extent to which this power is misused. When you receive a hard copy letter of recommendation in the mail—for a tenure case, say—you probably do not make 50 photocopies of the letter and send them off to 50 different mathematicians. First, such an action would be rude; second, it could have legal repercussions. For a written letter, the sender owns the contents and the recipient owns the piece of paper and that particular *form* of its contents (that is the law). A similar legal protocol has been proposed for *e*-mail, although at this writing the legislation has not been approved. What I am discussing here is not so much the law as common sense and common courtesy.

  People forward *e*-mail all over the place, with hardly a thought for the consequences. The courteous thing to do is to ask the author before you forward anything. Many people send me *e*-mail messages that say "Please delete this message after you have read it" (the implicit message here is "Don't forward this to anyone!"). I am always punctilious about adhering to such a request, and I hope that others are similarly considerate of my requests for discretion.

- Electronic mail is not as secure as other forms of communication. Any superuser on your system can eavesdrop on your *e*-mail, and computer bandits can break into the system and perform all sorts of nasty deeds. Thus you need to exercise some restraint with respect to what you say over *e*-mail. Many of us use hard copy letters and the telephone for the most delicate matters.

- In the early days of *e*-mail, a user had to be careful of line length: lines longer than 80 characters were often truncated by either the sending or the receiving editor. Given that some characters could be added in transit, it was best to keep lines to 72 characters in length. Most editors and mail spoolers now can handle longer lines, but careful users still keep lines no longer than 72 characters.

Some mail spoolers and *e*-mail editors insert line breaks into ASCII files that they receive. (The likelihood of this inconvenience increases if your lines are long.) Thus a perfectly good TEX command like \smallskip could be transmogrified to \smal at the end of one line and lskip at the start of the next line. If you are lucky, you will catch this glitch with a spell-checker. Of course you can bullet proof your file by UUENCODE-ing it before sending it.

If you send a file to a friend with lines that are longer than 80 characters, and if he endeavors to print it out cold, then the lines are likely to be chopped off in the hard copy. The industrious high-tech recipient will reformat each paragraph before printing— using <Esc>-q in EMACS or an analogous command on other systems. Other recipients will miss a lot of information.

Also avoid beginning any line with "From" or "from", as this word is proprietary to *e*-mail (and will result in unwanted characters being added to your document during the *e*-mail transmission process). For example, in order to protect the special use of "from", *e*-mail will replace it with ">from" when it occurs at the beginning of a line.

Electronic mail, or *e*-mail, is a marvelous tool. It has affected the mathematical infrastructure, and has altered the way that many of us collaborate and communicate. If each of us would exercise just a little *e*-mail etiquette, then the annoyances attendant to *e*-mail would be minimized.

Chapter 5
Books

Some books are to be tasted, others to be swallowed, and some few to be chewed and digested.

Francis Bacon
Essays [1625], Of Studies

No man but a blockhead ever wrote except for money.

Samuel Johnson
quoted in Boswell's *Life of Samuel Johnson*

I never think at all when I write
nobody can do two things at the same time
and do them both well.

Don Marquis

A writer and nothing else is a man alone in a room with the English language, trying to get human feelings right.

John K. Hutchens

The writer who loses his self-doubt, who gives way as he grows old to a sudden euphoria, to prolixity, should stop writing immediately: the time has come for him to lay aside his pen.

Colette

You can't polish cow chips.

paraphrased from Lyndon Johnson

5.1   What Constitutes a Good Book?

Mathematics books are written all the time. Go to the library and pull one at random off the shelf. Looks pristine, does it not? Or perhaps only the first fifty pages show signs of reading. Many an author lavishes all his enthusiasm and creativity and energy on the first part of his book; he then runs out of steam for the remainder. Unfortunately, it is the reader who suffers the consequences.

Writing a good book requires more effort than many authors are willing to give to the task. Writing a good *mathematics* book requires special insights and skills. In my view, the hard work is worth it. When you write a good mathematics paper, it is only read by a small group of people. But write a good book and a lot of people will see it. The book is a way of planting your flag, of putting your stamp on the subject, of sharing with the world the fruits of your hard labor.

My advice is not to consider writing a book until you have tenure and are established somewhere. The task is just too time consuming, and is often not construed as a positive contribution toward the tenure decision. Put differently, and a bit simplistically, the view of the world is that an Assistant Professor should be writing research papers and becoming established in the research community. Once you have done that, and achieved tenure status, then you have the leisure to consider other pursuits.

Now let us consider what makes for a good book. First, and foremost, you must have something to say. If you are only repeating, or paraphrasing, what has been said before then you are contributing nothing to the subject. Second, you must have a plan for saying it. The best method for writing a book is to immerse yourself thoroughly in the subject. The book itself becomes your "world" for a couple of years. A place to begin is to write a detailed outline of the book. Begin by writing chapter headings. Then fill in some section headings. After a while, the juices begin to flow and you will find that you cannot write fast enough to keep up with the outline developing in your head.

Once the book outline is written, it should be emblazoned on your frontal lobes. Carry it with you (in your head) all day long. I find, when writing, that I am constantly jotting down thoughts or topics or phrases that occur to me throughout the day. These can arise in conversation, or in lectures, or while daydreaming. If you are thoroughly involved with the project, then they come up.

Once you have a detailed plan of what you are going to do (and you are not bound to this plan, for it will evolve as your work unfolds), then you should begin to write. Write a chapter at a time. Completely immerse yourself in each chapter. If, while writing Chapter 3, a thought occurs to you about Chapter 6, then make a note. You can, especially in the computer environment, jump from one chapter to another. But the process can become confusing. Safest is to make a note—in a notebook perhaps, or in a computer file that you can pull up instantly. Then, when you begin work on Chapter 6, you have all your notes to work from.

Remember, as you write, that you are taking material that you have thoroughly digested and internalized and are presenting it to your readers—many of whom are tyros. Thus you must perform a reverse evolution to put yourself in the shoes of the learner. This may be hard to do at first, but it is a worthwhile exercise: it helps you to see as a whole how the subject is built and what questions it answers. It helps you to understand motivation and foundations.

Keep in mind that organization is a powerful tool. I have seen too many math books that state lemmas parenthetically. Here is an example:

> We thus see that every pseudo-melange is a hyper-melange. (We use here the fact that every pseudo-melange is complete. *Proof:* Let $\mathcal{M}$ be a pseudo-melange. Calculate its first Sununu cohomology group, etc.) ✠

Here the author is writing a love letter to himself. If you write such an epistle, then few will read it and fewer still will derive anything from it. Especially when writing with a computer, you can always add a lemma—wherever it is needed—and add suitable connecting material as well. Do not succumb to the temptation to skip this part of the writing regimen. Most of the process of developing a book consists of attending to details like making sure that all your lemmas and definitions are in place before you need them. You *must* attend to these matters.

To recast what I have been trying to say in the last few paragraphs, the first blush of writing can be lots of fun. You organize a subject in your head, or on paper. In a flurry of enthusiasm, you write a draft on paper. You see the subject begin to shape up as you, and only you, see it. You begin to take possession of this circle of ideas. The process is exciting and stimulating.

But then the moment of truth arrives. If you want to turn this random sequence of meditations into a publishable book, one that people will *read,* then some hard work lies ahead. You must go through the MS line by line, detail by detail, attending to context, syntax, logic, motivation, and many other details as well. You will proofread the same passages over and over again. Frequently, you will have to swallow your pride and rewrite an entire section, or reorganize an entire chapter. The revision process is hard, tedious work and not for the faint of heart.

You must put yourself in the shoes of the first year graduate student, or whoever represents the ground floor of those who might read your book. Where will such a reader get hung up, and why? What can you, as the author, do to help this person along?

Finding an original way to develop the proof of the latest theorem in your subject is always a pleasure. Reorganizing that material in a new way, for your six or eight close buddies in the field, is rewarding. Much less stimulating is writing a chapter of motivation and background material. But, thinking in terms of the longevity and impact of your book, you must learn to admit that both of these tasks are of paramount importance. The latter is not going to have people buying you drinks at the next conference, but it will help your book to have an impact on the infrastructure of your subject.

To summarize, what makes for the writing of a good book is hard work and unstinting attention to detail. Frequently the work required is tedious, and you will ask yourself why you cannot assign it to a secretary or a graduate student. The answer is that you are producing *your book,* and it is for the ages, and you want it to come out right.

5.2   How to Plan a Book

The business of planning a book has been touched on in the previous section. Here we flesh it out a bit.

A common way to develop a mathematics book is first to teach a course in the subject area. Indeed, teach it several times. Develop detailed notes for the course. Polish them as you go. Get your students and colleagues to read them, annotate them, criticize them. Become a good observer: note which parts of the book make sense to your audience and which require additional explanation from you. Use these notes and observations as a take-off point for the book.

Mathematicians appear to be a shy, introspective lot.[1] It seems to exhibit too much hubris for a mathematician to say "Now I shall write a book on thus and such." More often than not, the mathematician sneaks into the task; and a good way to do this is to develop lecture notes.

This lecture notes approach has several advantages over writing the book cold. First, you have the opportunity to classroom test the material, to see in real time how students react to it, and to modify it according to what you learn from the experience. Second, when you teach a course you are completely involved in the material, and it is natural to develop it and revise it as you go. Third, you can show your lecture notes to colleagues—without much fear of embarrassment because, after all, they are only lecture notes—and learn from their comments and criticisms. Fourth, if the material does not seem to be developing expeditiously, you can abandon the project without losing face. After all, these were only lecture notes.

It also helps to have a collaborator. Imagine going to a colleague at a conference or other group activity and saying "You know, there ought to be a book on *badeboop badebeep*." If the colleague indicates assent, then you can begin to describe what material ought to be in the book. Before long, you are swapping ideas, building each other's enthusiasm. Soon enough, you are writing a book together. Your collaborator is a reality check, and reassures you that you have not set for yourself a long-term fool's errand (for example, it would certainly be the pits to find out after two years of hard work that your book topic "Generalized

---

[1] An introverted mathematician is one who looks at his shoes when he talks to you. An extroverted mathematician is one who looks at *your* shoes when he talks to you.

Theory of Fluxions and Fluents" was no longer a matter of current interest).

Of course writing something as big as a book with a collaborator has its down side too. There will be periods when you are raring to go and he is busy getting a divorce, or learning to chant "Na myoho rengae kyo", or moving into a yurt. Or conversely. Taking on a book collaborator is like adding a member to your family. And the family could become dysfunctional.

I want to leave you with one important thought about planning a book. Try to have the entire vision of the book in place before you launch full steam into the project. Such planning enables you to keep your sense of perspective, to know how much has been accomplished and how much remains to be done. It also helps to prevent you from wandering off onto detours, or from developing specious lines of investigation. I have written books where I have just started writing and let the course of events dictate where my thoughts would lead me. Sometimes this worked well; more often it did not. After writing many books, I can say with some confidence that the planned approach is far superior.

5.3   The Importance of the Preface

I have already indicated in Section 3.4 why the Preface to any project is an important feature. For something as grandiose as a book, the Preface is paramount. Writing the Preface is part of the planning process, and it acts as your touchstone as you develop the project.

Indeed, while I am writing a book I often take a break and spend some time staring at my Preface and my Table of Contents (TOC). It may well be that, at an advanced stage of the writing, I no longer agree in detail with what the Preface and TOC say. But when I wrote the Preface and TOC my thoughts were organized and galvanized and I knew exactly what I was trying to accomplish. Studying the Preface and TOC is a way of reorienting myself.

And remember that your reviewers and your readers, if they are smart, will study your Preface and TOC in detail. These two essential front matter items will give them a preview of what they are about to read, and how to go about reading it. Just as you write the introduction to a research paper with the referee in mind, endeavoring to answer or at least minimize all his observations and objections, so you write your Preface and TOC with a view that you are deflecting all the reader's *But*'s.

Your Preface should not spare any detail. You have obviously thought about why existing books do not address or fill the need that your book fills. Spell this out in the Preface. You have thought about why your book has just the right level of detail and the right prerequisites. Say this in the Preface. You have thought about why your point of view is just the right one, and the points of view in other books are either outdated or misguided. Say so (diplomatically) in the Preface.

Even if you were to write your Preface, polish it to perfection, and then put it in the paper shredder, it would have been an important and worthwhile exercise to write it. Writing the Preface is your (formal) way of working out exactly what you wish to accomplish with your book.

5.4 The Table of Contents

In some sense, there is no way that you can know what will be in your book before you have written it. But you certainly will know the milestones, and the big ideas. In writing a novel, it may be possible to begin with "It was the best of times, it was the worst of times ..." and then let the ideas flow; however, technical writing demands more deliberation. Somehow, writing "Let $\epsilon > 0$" does not set one sailing into a disquisition on analysis. Mathematics is just too technical and too complex; you must plan ahead.

Writing the TOC is part of the early process of developing your book. It may hurt at first, and it may not feel like fun. But you will launch into writing Chapter 1 more easily if you know in advance where you are headed; conversely, if you do not know where you are headed then how can you possibly begin? Treat the writing of the TOC like working out on your NordicTrack®: just do it.

Make the TOC as detailed as you can. The more thoroughly that you can map out each chapter and each section, the more robust your confidence will become. That is, it will be much clearer that you can and will write this book. Always remember as you supply details that you are not wedded to this particular form of the TOC. You can, and no doubt will, change it later.

If you find yourself unable to write the TOC, then maybe God is trying to tell you something. Maybe you were not cut out to write this book or, worse, maybe you have nothing to say. Writing the TOC is an acid test. You will have to write it eventually. What makes you think that you will be able to write it *after* having written all the chapters if you cannot write it before? Does this make any sense? Write it now.

5.5 Technical Aspects: The Bibliography, the Index, Appendices, etc.

If you write your book using LaTeX, or using the macros included with the book [SK], then you have a number of powerful tools at your disposal for completing some of the dreary tasks essential to producing a good book.

In the old days, when an author created the index for a book, he proceeded as follows. (For effect, let me paint the whole dreary picture from soup to nuts.) First, the author sent his manuscript into the publisher. For a time, he would hear nothing while the copy editor was working his voodoo on the MS. Then the publisher sent the author the copy-edited manuscript. This gave him the opportunity to reply to the editor's comments and suggestions. For example, the editor might have changed all the author's *that*'s to *which*'s or vice versa. The copy editor might have said "You cannot call $G(x, y)$ 'the Green's function' because that is ungrammatical." Or "you cannot refer to 'Riemannian metrics' in Chapter 10 because, when Riemann's name came up in earlier chapters, it was not in adjectival form." (Both of these have happened to me; in the penultimate example, I was advised to call $G(x, y)$ "the function of Mr. Green".) In any event, the author slugged his way through the manuscript and made his peace with the copy editor, sometimes via a shouting match over the telephone.

At the next stage the author received "galley proofs". These were printouts of the typeset manuscript, but not broken for pages. Galley proofs were often printed on paper that was 14 inches long or more. The author was supposed to read the galleys with painstaking care, paying full attention to all details. The main purpose of this proofreading was to weed out any errors—mathematical or linguistic or formatting or some other—that were introduced by the typesetter. At the next, and final, stage the author was sent "page proofs". Now the author was receiving his manuscript broken up into pages, and appearing more or less as it would in the final book. Space had been made for figures, and the pages had running heads and actual page numbers. At this propitious moment, the author was (at least in theory) no longer checking for mathematical, English, or typesetting errors. In the best of all possible worlds, at this stage a check was being made that the page breaks did not alter the sense of the text, nor did they result in figures being misplaced.

And it was at the page proof stage that the author made up the index. First, he went through the page proofs and wrote each word to appear in the index on a separate 3×5 card, together with the correct page reference (which was only *just now* available, since the author was working for the first time with page proofs). Then the author alphabetized all the 3×5 cards. Finally, the author typed up a draft of the index.

In the modern, computer-driven environment for producing a book, the production process is considerably streamlined. If the manuscript is submitted to the publisher in some form of TeX, then usually the copyedited manuscript stage and also the galley proof stage are skipped. The author works with page proofs only, and that is his last "pass" over the manuscript. The entire business of writing words and page numbers on index cards, alphabetizing them, and then typing up an index script is gone. Here is the new methodology:

Imagine, for example, that you are using LaTeX. You can go through your ASCII source file and tag words. (*You can do this at any stage of your writing—indeed, you may do it rather naturally "on the fly" while you are creating the book.*) For instance, suppose that somewhere in your source file the word "compact" occurs, it is the first occurrence, and you want that word to be in the index with that particular page reference. Then you put the code `\index{compact}` immediately adjacent (with no intervening space) to the occurrence in the text of the word "compact"; thus `\index{compact}` now appears in your TEX source file. This additional TEX code does not change the printed TEX output. But the indexing commands cause all the words that have been marked for the index to appear in a single file, called MYFILE.IDX (assuming that the original file was MYFILE.TEX), together with the appropriate page references—*after* you have compiled the source file. You can then use the UNIX command `makeindex` to alphabetize the file MYFILE.IDX and to remove redundancies. The procedure is documented in the LATEX book [Lam], or in the file MAKEINDEX.TEX. (Alternatively, you can use operating system commands to alphabetize the index, and then do a little editing by hand to eliminate repetitions and redundancies. The entire process usually takes just a few hours.) The disc that is included with the book [SK] also includes macros that will assist in the making of an index.

The reference [SG, pp. 76-96] treats all the technical aspects of compiling a good index. The book [Lam] has a nice discussion of the notion that you should index by *concept,* not by word. The former method allows the reader to find what he is looking for quickly; the latter adds—unnecessarily—to the reader's labor. A good, and thorough, index adds immeasurably to the usefulness of a book. My claim is particularly true if your book is one to which a typical reader will refer frequently and repeatedly—for example if your book is meant to be a standard treatment of a mathematical field. Many otherwise fine mathematics books are flawed by lack of an adequate index (or, for that matter, lack of an adequate bibliography).

There are professional indexers who can produce a workable index for any book. But nobody knows your book better than you, the author. *You* should create the index. Given that modern software makes the creation of an index relatively painless, there really is no excuse for not creating one yourself.

LATEX can also compile a TOC for you. You will have little difficulty making a TOC by hand—you merely need to copy all chapter and section titles to a single file and verify the page numbers, and then wrestle with the formatting. LATEX executes all these formatting and tabulating operations for you automatically, while you are compiling the book for printout. It is that simple.

Similar comments may be made about the Bibliography—this procedure has already been discussed in detail in Section 2.5. The book [SK] tells you how to write TEX macros to compile a glossary, a table of notation, or any similar compendium. The process is rather technical, and I shall not describe it here.

I conclude with a few words about Appendices. You will sometimes come to a point in your book where you feel that there is a calculation or a set of lemmas that you know, deep down, must be included in the book; but it will be painful to write them, and they will interrupt the flow of your ideas. The solution then is to include them in an Appendix. Just say in the text that, in order not to interrupt the train of thought, you include details in Appendix III. Then you state the result that you need and move on. This practice is smart exposition and smart mathematics as well. It is also a way of managing your own psyche: when you are attempting to tame technical material in the context of your book proper, then it becomes a burden; if instead you isolate the same material in an Appendix, then you loosen your fetters and the task becomes much easier.

An Appendix also could include background results from undergraduate mathematics, alternative approaches to certain parts of the material, or just ancillary results that are important but too technical to include in the text proper. Appendices are a simple but important writing device. Learn to use them effectively.

5.6 How to Manage Your Time When Writing a Book

Many a mathematics book is started with a bang, two-thirds of it is written, the writer becomes bogged down in a struggle with a piece of the exposition, or the development of a particular theorem, and the book is never completed. I cannot tell you how often this happens; perhaps more frequently than the happy conclusion of the book sailing to fruition. I imagine that the same hangup can occur for the novelist, or for the historical writer.

I would be naive, indeed silly, to suggest that those who cannot complete their books are just insufficiently organized. Or that such people have not read and digested my advice. Anyone can develop writer's block, or can arrive at a point where the ideas being developed just do not work out, or can just lose heart. We as mathematicians, however, are accustomed to this dilemma. Most of the time, when we write a paper, things do not work out as we anticipated. The hypotheses need to be adjusted, the conclusions weakened, the definitions redeveloped. If you are going to write a book then you will have to apply the same talents in the large. But you also need to think ahead to where the difficulties will lie and how you will deal with them. One of the advantages of doing mathematics is that nothing lies hidden. We can think and plan the entire project through, if only we choose to do so.

People in twelve-step programs, with chemical dependencies, are taught to live one day at a time. Such people are taught to concentrate on the "now". If you are writing a book then, on the one hand, you cannot afford this sort of shortsightedness. You must plan ahead, and have the entire project clearly in view. If you kid yourself about how Chapter 8 is going to work out then, when you get to Chapter 8, you are going to pay. By analogy, if you write a paper in such a fashion that you shovel all the difficult ideas into Lemma 3 then, when it comes time to write and prove Lemma 3, you must face the music. You cannot fool Mother Nature.

But, having said this, and having (I hope) convinced you of the value of planning, let me now put forth the advantages of tunnel vision. Once you have done the detailed planning, and you are convinced that the book is going to work, then develop an extremely narrow focus. Pick a section and write it. You need not write the sections in logical order (though there is some sense to that). But, once you have picked a section to work on, then focus on that one small task, that one small section, and do it. If some worry about another section, or another chapter, crops up then make a note of it and then press ahead with the writing of your chosen section. Bouncing around from section to section, and chapter to chapter—chasing corrections around a never-ending vortex—is a sure path to disillusionment, depression, and ultimate failure. You can always set up scenarios for defeat. Your book-writing project can turn into a black hole, both for your time and for your psychic energy. Writing a book is a huge task; nobody will blame you if you give up, or abandon the effort. But with some careful planning, with an incremental program for progress, and with some stamina, you can make it to the end.

Paul Halmos [Ste] advocates, and describes in detail, the "spiral method" for writing a book (or a paper, for that matter). The idea is this: first you write Chapter 1, and then move on to Chapter 2. After you have written Chapter 2, you realize that Chapter 1 must be rewritten. You perform that rewrite, re-examine Chapter 2, and then you move on to Chapter 3, after which you realize that Chapters 1 and 2 must be rewritten. And so forth. If you are writing by hand, with a pen on paper, then the spiral method takes place in discrete steps as indicated. If, instead, you write with a computer then the spiral method can take place in a more organic fashion: as you are writing Chapter 3, and realizing that Chapter 1 needs modification, you pull up Chapter 1 in another window and begin to make changes while you are thinking about them. If those changes in turn necessitate a massage of Chapter 2, then you pull it up in a third window. The advantage of doing things in discrete steps, as described by Halmos, is that you always know where you are and what you are doing; the disadvantage of the organic approach is that you can become lost in a vortex—caroming around among several chapters. The technique must be used with care.

It can only improve your work to review Chapters 1 through $(n-1)$ after you have written Chapter n. On the other hand, if you do use the out-of-the-box spiral method, as described and recommended by Halmos, then one upshot will be that Chapter 1 of your book will receive more attention than any other part, Chapter 2 will receive the second greatest dose of attention, and so forth (for the proof, use induction). As a result, your book *could* appear to the reader to become looser and looser as it proceeds. Perhaps this is an acceptable outcome, for only the die-hards will get to the end anyway. But when you adopt a method for its good points, also be aware of its side effects.

Certainly choose a method that works for you—organic, inorganic, spiral, or some other—and be sure to use it. If there is any time when it is appropriate to be organized, methodical, indeed compulsive, that time is when you are writing your book.

No matter what method you adopt for reviewing and modifying your work, keep this in mind: only wimps revise their manuscripts; great authors throw their work in the trash and rewrite. Such advice causes many to say "That is why I could never write a book; it is sufficient agony just to write a short paper." Rewriting is not so difficult; in many ways it is easier than figuring out where to insert words or to substitute passages. Treat your first try as just getting the words out, for examination and consideration. Once the thoughts are lined up in your head, then the first draft has served its purpose; you may as well discard it (and *don't peek!*). The next go is your opportunity to shape and craft the ideas so that they sing. The next round after that allows you to polish the ideas so that they are compelling and forceful. The final step allows you to buff them to a high sheen.

Use the advice of the last paragraph along with a dose of common sense. After you have struggled for a month to write down the proof of a difficult proposition, you are not going to throw it in the trash and start again. My advice here, as throughout this book, applies selectively.

Once you have arrived at (what appears to be) the end of the task of writing your book, you still are not finished. There remains a lot of detail work. You must prepare a good bibliography (Sections 2.6, 5.5). You must prepare a good, detailed, index (the computer can help a lot here—see Section 5.5). If appropriate, you should prepare a Table of Notation. You might consider building a Glossary. None of these tasks is a great deal of fun. But they will increase the value of your book immeasurably. They can make the difference between an advanced tract accessible to just a few specialists, or a book that opens up a field.

5.7   What to Do with the Book Once It Is Written

You have written your *magnum opus*, slaved over it for two or more years, shown it to colleagues, received the praise of student and mentor alike. The manuscript is now polished to perfection. There is no room for improvement. Now what do you do with it?

The rules for submitting a book manuscript to a publisher are different from those for submitting a research paper to a journal. The hard and fast rule for the latter is that you can only submit a research paper to one journal at a time. Most research journals tell you up front that, by submitting a paper, you are representing that it has not been submitted elsewhere.

Not so for books. You can submit a book manuscript simultaneously to several different publishers. These days there are just a few mathematics publishers—especially for advanced books. Get a feel for the different publishers by looking at their book lists. You will see what quality of books and authors they publish, and in what subject areas. Some publishers, such as the AMS, CRC Press, Springer-Verlag, and Birkhäuser, have several book series in mathematics. Familiarize yourself with all of them so that you can make an informed choice. Talk to experienced authors to obtain the sort of information that cannot be had from advertising copy.

If you want to jump-start the publication process, then you can begin long before your book is completed. For example, if you are looking for a typing grant or an advance, then you may wish to begin negotiations with publishers after you have written just two or three chapters. Submit them, along with a Preface or Prospectus[2] (the marketing version of a Preface) and a TOC. And of course include a brief cover letter saying who you are, what book you are writing, and exactly what materials you are remitting.

Always send a manuscript to a publisher by either registered or certified mail—return receipt requested. There are both practical and religious reasons for doing so. First, it requires some effort and expense to prepare a manuscript, plus the figures, plus the discs, for submission to a publisher. You want to protect your investment of time and money; so special mail services and even insurance are definitely in order. Less obvious is an artifact of the way that publishing houses work: items that

---

[2]Like a Preface, the Prospectus will describe what the book is about and why you have written it. Unlike a Preface, the Prospectus will describe the audience, the competing texts, the types of courses that could use the book, and the types of schools and departments that might adopt the book.

arrive by regular post tend to get thrown into a pile; items that arrive by registered or certified or express mail are given special treatment. Stop and think about how many manuscripts, or how many pieces of mail, a big publishing house will receive in any given business day. Now you will understand why you should take pains to ensure that your manuscript receives the particular attention that it deserves.

In order to be able to negotiate intelligently with a publisher, be sure to have the following information about your book under control:

1. Subject matter and working title

2. Level (graduate, undergraduate, professional, etc.)

3. Classes in which the book could be used

4. Existing books with which your book competes

5. Working length

6. Expected date of completion

The publisher needs to know a subject area and working title for in-house and developmental purposes. The guys in the suits refer, among themselves, to the "Krantz project on fractals". So they need a working title. They need to know a working length and a sketch of the potential market so that they can price out the project. They need to know an approximate due date so that they can deal with scheduling (a non-trivial matter at a publishing house).

I am the consulting editor for a book series. One of my earliest authors completed his book two years late, with a book twice the length originally projected; also the book was on a different subject than that contracted, and with a different title. I cannot tell you how much trouble I had persuading the publisher to go ahead with the project. When you are dealing with a publishing house you are dealing with business people. You must endeavor to conform to their view of the world.

If the publisher is interested in your project, then he will probably solicit reviews. Some publishers will ask you to suggest reviewers for your project. Most will not. Expect the reviewing process to take three or four months. Expect to see two to four reviews of your work.

One of the most difficult, and valuable, lessons that I have learned as an author is to read reviews. By this I mean to read them intensely and dispassionately and to learn what I can from them. Forget reacting to the criticisms. Forget justifying yourself. Forget answering the reviewers' comments. The point is this: even if you cannot understand what the reviewer is thinking, what he describes is nevertheless what he saw when reading the manuscript. The review describes the impression that the manuscript made on him. The main question you should be asking yourself as you read the reviews is "What can I learn from these reviews?" "How can I use these comments to improve my book?" There is generally something of value in even the most negative of reviews.

Usually the publisher has established an initial interest in your project by looking at your Prospectus and TOC, and by agreeing to undertake the cost of reviewing. If the consensus of the reviews is favorable, then the publisher will most likely decide to publish your book. He will then ask you to take the reviews under advisement, and only that. The editor may want to discuss them with you, and may even want your detailed reaction to them. But few, if any, publishers will hold you accountable for each comment made by each reviewer.[3]

Remember this! And what I am about to say applies to research papers and to books and to anything else that you submit for review: the reviewer is not responsible for the accuracy and correctness of your work. There is only one person who bears the ultimate responsibility, and that is you. Many reviewers will do a light reading, or an overview, or will read the manuscript piecemeal, according to what interests them. If the reviewers give you a "pass", then that is good. But this "pass" not a benediction, nor even a suggestion that everything you have written is correct. You must check every word, and you yourself must certify every word.

---

[3]Note that these remarks do not apply to the writing of a book at the lower division level, for the so-called "College Market". Such a project is more of a team effort: you and the reviewers write the book together, in a sort of Byzantine tug-of-war procedure. The process is best learned by consenting adults in private, and I shall say nothing more about it here.

In any event, the period immediately following the review process is your chance to take a couple of months and polish your manuscript yet again. (You will also have the opportunity to make small changes later on in the page proofs. But the post-review period is your last chance for substantial rewriting.) *Treat this as a gift.* It would be embarrassing to publish your book blind—with no reviews—and then to have your friends point out all your errors and omissions, or (worse) that your point of view is all wrong. The reviewing process, though not perfect, is a chance to collect some feedback without losing face and without any repercussions.

After you have polished your MS to your satisfaction, and presumably shown it to some friends and students and colleagues, then you submit the final, polished draft to the publisher. Many publishers will want this manuscript to be double or triple spaced, so that the various copy editors and typesetters will have room for their markings and queries. The TEX command \openup$k$ \jot, where k is a positive integer, will increase the between-line spacing in your TEX output by an amount proportional to k.

Now here is one of the great myths that exists at large in the mathematical community. People think that, in 1996, you send a diskette, with TEX code on it, to the publisher. The publisher puts the diskette in one end of a big machine and a box of books comes out the other. Technologically this phenomenon is actually possible. But a top-notch publishing house has a much more exacting procedure.

Here, instead, is what a good publishing house does with your manuscript and disc. First, an editor decides whether your book is ready to go into production. He may show your "final manuscript" to a member of his editorial board, or he may make the decision on his own. But this hurdle must be jumped. Once the book goes into production, some copy editing will be done. The actual amount will vary from publishing house to publishing house. During the copy editing process, your spelling, grammar, syntax, consistency of style, and other nonmathematical aspects of your writing will be checked. Depending on the density of corrections at this stage, you may or may not be contacted. You may have to submit another manuscript.

Then your project goes to a typesetter. Yes, truly, it goes to a typesetter. But what, you ask, about TEX? I thought that when I used TeX then *I* did the typesetting. To a degree, that is true. But the publishing industry has extremely exacting standards. In a book produced to the highest quality demands, the space above and below each theorem should be the same. All displayed equations should be formatted in the same way. Left and right page bottoms should align (this last task is something at which Plain TeX does not excel; LaTeX handles the issue with the \flushbottom command). No page should begin with the last two words of a paragraph, nor should any page end with the first two words of a paragraph (these stragglers are called "widows"). Many other strictures apply as well. A trained professional, with an eye much more demanding than yours or mine, must go through the TeX file, with your manuscript in front of him (so that he can be sure of how the output is supposed to appear) and make many corrections. Space must be made for figures. Running heads must be verified. There are numerous minute details that *someone* with technical training must check.

If your project were typeset the old-fashioned way, with movable type—say that it has 400 pages—then the typesetting job would cost $15,000–$20,000. If instead you produce a TeX file to a level of reasonable competence, then the adjustments that I described in the last paragraph will cost $5,000 to $7,000. So TeX *does* save money in the publishing process.

After the TeXnical typesetter massages your manuscript, page proofs are generated (these days, the galley proof stage—see Section 5.5—is often skipped). The page proofs are sent to you, the author, for proofreading. The proofs that you are sent may have "author queries" penciled in the margins—"Is this formula supposed to appear this way?", "Do you indeed mean to say this?", etc.—and you must answer them all. This moment is also, for all practical purposes, your last check of the book. You can make minor corrections at this time. If you must insert paragraphs, rearrange material, or do other editing that will affect page breaks, pagination, and the lengths of chapters, then the publisher might fuss. If your revisions are extensive (more than 10% or 15%) then the publisher might charge you. At the time that you receive the galley proofs, you will also probably receive proofs of the figures. Usually, each figure is on a separate sheet. The space allocated for each figure is shown in the page proofs, and each is labeled. The legends and labels for the figures will be provided as streams of type on a separate page; these are easy to overlook if you do not know what they are. Check that each figure appears as it should, that it appears in the right place, and that the legend and label are correct. (The proofs of figures that you will see may be "larger than life". Not to worry: the publisher shrinks them to fit.)

After you have approved the page proofs (often you are asked to sign off on each chapter, certifying that you have checked everything and that the present form is the way that you want things to appear in the printed book), then that is the end of your role in the publishing process (but see the *caveat* below about the dreaded Marketing Questionnaire). I suggest that you *insist* on seeing the title page before the book goes to press. It happens—not often—that an author's name is misspelled or an affiliation is rendered incorrectly. Such an eventuality is embarrassing for everyone. It is best to avert it.

Even though your role is at an end, let me say a few words about what happens next. As is mentioned elsewhere in this book, when a TeX file, consisting of `ASCII` code, is compiled then the result is a DVI file. Typically, this "Device Independent File" is then translated, using software and without human intervention, to a `PostScript`® file. Why `PostScript`®? Many high resolution printers such as a Line-a-Tron 300®—with anywhere from 1200 DPI (dots per inch) up to 2400 or more DPI—read `PostScript`®. Once the files for the book have been translated into `PostScript`®, then the book is printed out at high resolution on RC (resin coated) paper. The result is a reproduction copy (or *repro copy*) of your book printed on glossy, nonabsorbing paper, at extremely high resolution. All the smallest subscripts and superscripts will be sharp and clear, even under magnification.

At this stage some "composition" may be necessary. Even though we all like to revel in the notion of inputting graphics formatted in `PostScript`® into our TeX files (see Section 6.7), it happens that the graphics in a book are (at least today) often not handled in that fashion. You will frequently find it more convenient to insert the graphics (which have been printed out elsewhere, using different software) by cutting and pasting.

The repro copy of the book is then "shot". Here, to be "shot" means to be photographed. The pages of the book are photographed onto film, in the fashion familiar to anyone who takes snapshots. But it is not printed onto photographic paper (what would be the point of that?—it is *already* on paper). Instead, the negative is then exposed or "burned" into chemically treated plates. These plates are the masters from which your book is printed. (This process is becoming ever more streamlined. Today at the AMS, the "repro copy" step is skipped altogether; the production department goes directly from the electronic file to the negative.)

Once the printing, or lithographic, plates are prepared, then the rest of the printing process—printing, cutting, and binding—is quite automated. Good books are printed sixteen pages to a sheet, and then folded and cut. This procedure results in the "signatures" that you can see in the binding of any high-quality book (not a cheap paperback). (In the old days the publisher did not cut the signatures; a serious reader owned a book knife, and did the cutting himself.)

Interestingly, the physical cost of producing a book—that is the printing, the binding, the cost of the paper—is well under $5 per volume; at least this is true if the print run is reasonably large. The difference in cost between producing a paperback volume and a hardback volume is about $1, depending on the quality of the papers used. So why do math books cost so much?

The pricing question for books is all a matter of marketing. To be fair, the publishing house has overhead. You remember the $5,000 to $7,000 for the services of a TEXnician? That is a cost that anyone can understand. Then the salaries of the editor, the publisher, the company president, the people in the production department, the costs of marketing, the physical plant, and so forth must come out of money earned from the sale of books. Most people, indeed most authors, are not cognizant of the cost of warehousing books in a serviceable manner (so that the books are readily accessible when an order comes in). Warehousing is a fixed cost that adds noticeably to the expense of each and every book that we buy. These last costs are called "overhead" or "plant costs", and play much the same role as the overhead for an NSF Grant. Most publishing houses figure the cost of producing a book by taking the up front, identifiable costs—technical typesetting, any advance to the author, print costs (often the printing is done by an outside firm), copy editing, composition, shooting—and then adding a fixed percentage (from 30% to 50%) to cover the overhead that was described above.

Then the editor does a simple arithmetic problem. He must make a credible, conservative estimate as to how many copies your book will sell in the first couple of years. Thirty years ago this was easy, since many libraries had standing orders for all the major book series. (For example, in the late 1960's, a company like Springer-Verlag or John Wiley could *depend* on library sales of 1000-1200 copies for each book!) With inflation, cutbacks, and other stringencies, libraries now pick and choose each volume. Thus the editor must make an evaluation based on **(i)** whether the book is in a hot area, like dynamical systems or wavelets, **(ii)** whether people in disciplines outside mathematics (engineers, for example) will buy it, **(iii)** whether students will buy it, and **(iv)** whether the book can be used in any standard classes. Other factors that figure in are **(a)** Is this the first book in an important field? **(b)** Is there stiff competition from well-established books? **(c)** How much effort is the marketing department willing to put into promoting the book? (You may suppose that the marketing department will promote any book that the editorial department sends in. On a *pro forma* level they will. But there is a delicate dynamic between these two publishing house groups, and a constant push and pull. A good editor takes pains to generate enthusiasm among the marketing people for particular books.) Having evaluated these factors, the editor writes a proposal for how many volumes of your book the house can expect to sell within a reasonable length of time (a couple of years). Then he figures in the company's standard profit expectation. This gives rise to the wholesale price of the book.

As an example, suppose that you write a book on a fairly specialized area of partial differential equations. After an analysis of the foregoing kind, the editor determines that the book is sure to sell 500 copies in the first two years. The up front costs are $15,000. Add 50% for overhead and that makes $22,500. Add 20% for the company's standard profit margin and that brings the total to $27,000. The wholesale price of the book must, after sales of 500 copies, bring in that much money. (If a given editor has several books that fail to meet this simple criterion, then he is out of a job.) Now do the arithmetic. You will find that the wholesale price of this book must be $54 per volume. Thus a bookstore will probably sell it for at least $70 to $80. Now do you understand why mathematics books cost what they do?

Incidentally, if the difference in cost between producing a hardcover copy of a given book and a paperback copy of that same book is about $1, then what accounts for the large difference in cost between hardcover and paperback books? The answer, apart from marketing voodoo, is that the costs of producing the book tend to be covered by the sale of the hardcover version. Thus the publisher has considerable latitude in pricing the paperback edition. John Grisham novels stay in hardcover format for more than one year before the paperback edition is released; usually, the hardcover edition sells millions of copies. The production costs, and the huge advance that Grisham garners for each of his books, are well covered by the hardcover sales. Thus the publisher is ready to make real money when the paperback edition is released. He can be imaginative both in pricing and in production values—if the physical cost of producing a volume is $5-$10, then he can price it for as little as $7-$15 and expect to sell a great many copies. (Interestingly, the entire notion of mass market paperbacks was invented in the early 1950's by Mickey Spillane and his publishers—Dutton and Signet. By 1955, Spillane had written three of the five best selling books in history—and he had only written three books! By contrast, Margaret Mitchell's blockbuster *Gone with the Wind* [Mit] sold fewer than a million copies in its first two years—all in hardcover, of course. James Gleick's *Chaos* [Gle] has sold about the same.)

Back to math books. In the preceding discussion there was an important omission. How does the editor make the market determination that I described three paragraphs ago? He can always consult his editorial board and his trusted advisors. But let me reassure you that he will certainly study your Prospectus and Preface, and he will pay close attention to your *Marketing Questionnaire*.

The latter item bears some discussion. Whenever you write a book for a commercial publishing house, and often for a professional society or a university press, you will be sent a Marketing Questionnaire to complete. I hate to complete these things, and you will too. But you must do it. I have heard authors say "I'll just phone the editor and talk to him about this stuff." Sorry; that just will not do. You must complete the questionnaire, and carefully.

What is this mysterious object? First, the questionnaire is long— often 10 pages or more. Second, it asks a lot of embarrassing questions: What is your hometown newspaper? Which professional societies might be interested in your book? What are the ten strongest features of your book? What is the competition? Why is your book better? In which classes can your book be used? What is typical enrollment in those classes? How often are they taught?

As mathematicians, we are simply not comfortable fielding questions such as these. We do not think in these terms. But, if you have been attending to the message of this section, then you can see how an editor can use this information to help price out the book. So why can you not just go over this stuff on the phone with the editor? One reason is that the editor needs this information *in writing*—for the record, and to show that he is working from information that *you* provided, and for future reference. The other is that the questionnaire will be passed along to the marketing department for the development of advertising copy and marketing strategies for your book. Like it or not, the Marketing Questionnaire is important. Take an hour and fill it out carefully.

When I was developing my first book, and negotiating with my publisher, I asked the editor what I would be peeved about three years down the line. He told me that I would be unhappy about the size of the print run, and I would be unhappy with the advertising. Then he explained to me how the world works. First, think about the sales figures that I described above. And think about the fact that a business must pay a substantial inventory tax for stock on hand. Extra books sitting around are a liability. And today (with new printing technology) small print runs are not so terribly expensive as they were even ten years ago. So if the publisher thinks that your book will sell 500 copies in the first couple of years, then the first print run is likely to be only 750. When that stock starts to run low, another 750 can be generated easily. The money saved per unit with a print run of 1500 (as opposed to 750) is relatively small, and is sharply offset by storage costs and inventory tax.

And now a word about advertising. There is nothing that an author likes better than to open the *Notices* of the American Mathematical Society or the *Mathematical Intelligencer* or the *American Mathematical Monthly* and to see an ad for his book. Of course a full page ad is best (and almost never seen), but a half page, or quarter page, or even an ad shared with eleven other books, is just great. Typically, you will see such an ad just once for your book. After that, your name and the title of your book will appear in the company's catalogue. That is just the way it is. Marketing departments have convinced themselves that this point-and-shoot technique is the most effective way to market books. Many publishers rely on "card decks"—stacks of $3'' \times 5''$ cards, each with a plug for a single book—that are mailed in a block to mathematicians. Usually the potential buyer can mail in a card, without money, and receive a copy of a particular book for a 30 day examination period.

In the spirit of doing first things last, let me now say a few words about book contracts. When a publishing house is interested in publishing your book, then it will send you a contract. Typically, you will be offered a royalty rate of 10% to 15%. You will be given a deadline, and this deadline is definitely negotiable. Err on the conservative side (more time, rather than less), so that you have a fighting chance of finishing the book on time. If you do finish on schedule, then the publisher will take a shine to your project. If you do not, and the project is six months late, then most publishers will be forgiving; but, technically, a late project is no longer under contract!

A rough page length will be specified in the contract, and a working title given. Sometimes you will be offered an advance against royalties, or a typing grant. Sometimes you will be asked to certify that you will submit your manuscript in some form of TeX. Then there will be a lot of legal gobbledygook, most of which seems to be slanted in favor of the publisher. For the most part, it is. The publisher wants to be able to pull the plug on a project whenever and wherever it deems such an action suitable. Honorable publishers do not like to exercise this option, but they want to have the option available.

I can tell you that many authors—especially first-time authors—are quite uncomfortable with standard book contracts. This uneasiness stems, for the most part, from lack of familiarity. The details of the contract *can* be negotiated, and you should discuss with your editor any passages or provisions that you do not like. If the publisher wants *you* to render the artwork in final form, and you cannot or will not do it, then negotiate. If you do not like the deadline, then negotiate.

If the number of gratis copies of the work offered to the author is not adequate, then negotiate some more. Usually such negotiations are fairly pleasant. You will find the editor eager to cooperate—as long as your demands are within reason.

You may find it attractive to join the *Text and Academic Authors Association.*[4] This organization was formed to defend the rights of authors, and will help you in dealing with publishers. It also has a rather informative newsletter. And membership gives you access to a number of useful discounts, so that your dues are almost a wash.

I have dealt with many publishers. Most of them are very good to their authors (as well they should be) and most employ knowledgeable and competent editors. However, forewarned is forearmed. It is helpful to be familiar with the publication process before you launch into it.

---

[4] *Text and Academic Authors Association*, P. O. Box 160, Fountain City, Wisconsin 54629-0160.

Chapter 6
The Modern Writing Environment

Computers are useless. They can only give you answers.

Pablo Picasso

If he wrote it he could get rid of it. He had gotten rid of many things by writing them.

Ernest Hemingway
Winner take Nothing [1933]. Fathers and Sons

Easy reading is damned hard writing.

Nathaniel Hawthorne

In a very real sense, the writer writes in order to teach himself, to understand himself, to satisfy himself; the publishing of his ideas, though it brings gratification, is a curious anticlimax.

Alfred Kazin

On seeing a new piece of technology:

A science major says "Why does it work?"
An engineering major says "How does it work?"
An accounting major says "How much does it cost?"
A liberal arts major says "Do you want fries with that?"

Anon.

[With reference to Germany] One could almost believe that in this people there is a peculiar sense of life as a mathematical problem which is known to have no solution.

Isak Dinesen

6.1   Writing on a Computer

Writing on a computer is not for everyone. First, if you are going to write about a technical subject like mathematics, then you are going to have to learn TEX or another markup language or else you must learn the details of some technical word processing system. You may find, with considerable justification, that it is just agony to write the Cauchy integral formula on a computer—whereas you can dash it off in a jiffy when you write with a pen on paper. (Of course if you learn to use macros then you can write the Cauchy integral formula *more quickly*—not less quickly—with the computer. But computer writing is a *different* method of writing, and may not be for you.)

These thoughts are not trivial. If your writing tools interrupt your flow of thought then you are not using the proper tools, at least not the proper ones for *you*. If your writing environment is more of a hindrance than a help then you had better change it.

Clearly, when you are writing on a piece of paper with a pen or pencil, then you can easily and naturally jump from one part of the page to another. You can, in a comfortable and intuitive fashion, jot marginal notes and make insertions. You can put diacritical marks and editorial marks where appropriate. You can scan the current page, flip ahead or back to other pages, sit under a tree with your entire MS clutched in your fist, put Post-it® notes in propitious locations, tape addenda to pages, and so forth.

Now the fact is that almost all the "old-fashioned" devices described in the last paragraph have analogues in the computer setting. And the computer has capabilities that the traditional milieu lacks: magnificent search facilities, unbeatable cut and paste features, the power to open several different windows that either contain several different documents or several different parts of the same document, and many others as well. But you still must use the tools that work for you. If you have been writing with a pen on paper for many years, then you may be disinclined to change. At a prominent university on the east coast there is an eminent and prolific mathematician, who has access to any writing facilities that one might wish, and who writes by candlelight with a quill. That is his choice, and it certainly works for him. I also know people who take lecture notes, directly in TeX, on a notebook computer. This I cannot imagine, but it works for them.

In this section I want to say a few words about writing on the computer, and what I find advantageous about it. I will reserve comments about specific writing systems, like TeX, for a later section.

When writing on a computer, you can type as fast as you wish, never fearing for spelling or other errors. When you become acclimated to the medium, you can create text at least as fast as you would have with a pen (assuming that you know how to type), and the text will always be legible. You can make corrections, insertions, deletions, and move blocks of text with blissful ease. You can print out beautiful paper copy of your work (paper copy is called *hard copy*), and you can store your work on your hard disc or hard drive (also known as the *fixed disc*, or **C:** drive). You never need worry about misplacing all or part of your manuscript, since finding files on your hard drive is trivial. Even if, weeks or months later, the only thing you can remember about your document is a word or phrase in it, you will be able to find the file containing the document in seconds.

To illustrate this last point, I often find myself printing out another copy of a paper or chapter that I am working on, rather than trying to find where I put my last paper copy. I can find my file on my hard disc and print it out in just a moment; the old approach, more traditional and agonizing, of searching through my study for my hard copy could take hours. And remember this point: any tool that prevents your writing moods from being interrupted or jarred is a valuable one. My computer has eliminated, for me, the need to search my office for the paper copy that I want to work on. It saves me hours of time, and it saves me considerable irritation. Cherish those tools that make your life easier, and learn to use them well.

When working on a computer, you easily can keep every single version of a document you are writing. Suppose, for example that you are writing an article about diet fads among troglodytes. The first version of your article could be called **TROG.001**. After you modify it, the second version could be called **TROG.002**. The third would then be called **TROG.003**. And so forth. All these would be neatly stored, and accessible, on your hard disc. Compare with the situation, in a paper office, in which you had thirty-two different versions of a document. How would you store them all? How would you keep track of and differentiate among them? How would you access them? Note that a computer also assigns a time and date stamp to each file you process. Thus, when you do a directory reading, you would see something like this:

```
TROG  001      2357      9-21-94      11:15pm
TROG  002      3309      9-22-94       2:31pm
TROG  003      3944      9-24-94      10:42pm
TROG  004      4511      9-29-94       9:11am
TROG  005      3173      10-2-94       2:04am
```

We see here five versions of the paper. The third column shows the number of bytes in each version. The fourth shows the date on which the editing of that version was completed. The last column shows the exact time of completion.

Note that the versions grew in size until, in the wee hours of October 2, the author decided to discard more than 1300 bytes of the document; this resulted in version 005. Is it not reassuring to know that all the old versions are available, just in case the author decides to resuscitate one of his old turns of phrase? Whether or not you are in the habit of examining old drafts of your work, you will find it psychologically helpful to have all the old versions. When the work is complete you can, if you wish, discard all the drafts but the final one. But the fact is that mass storage space is so cheap and plentiful these days that every draft of every one of your works, even if you are Stephen King and Tom Clancy rolled into one, will only take up a small fraction of your hard disc.

If you do your writing with a first-rate text editor, as I do, then you have powerful tools at your disposal (see [SK] for a discussion of text editors). You can open several files simultaneously, have several different portions of the same file open at the same time, and have a bibliographic resource file open; with an environment like Windows95®, you also can have a CD-ROM thesaurus and dictionary open and be connected to the Internet—and you can jump from one setting to the other effortlessly. Given that any trip to the dictionary could take ten minutes the old-fashioned way, and ten seconds the electronic way, think of how much time you will save over a period of several years. Again—and here is *the* most important point—by using technology you circumvent the danger of your thought processes and your creative juices being interrupted.

Even though I am addicted to writing with a computer, hard copy plays an important role in my writing process. For, after I have written a draft, I print it out, lounge in my most comfortable chair, and proofread and edit. There exist methods of proofreading and editing directly on the computer—I shall not go into them here. But, because of my age and my training, I find that there is nothing like a paper copy and a red pen to stimulate critical thinking. You will have to decide for yourself what works for you.

There is a down side to writing with a computer; you can work your way past this one, but you had best know about it in advance. When you create a document on a computer system—especially if you use a sophisticated computer typesetting system like TEX—then the printed copy looks like a finished product. This makes it even more difficult than usual for you, the author, to see the flaws that are present. Even with handwritten copy you will have difficulty seeing that certain paragraphs must go and others must be rearranged or rewritten. But, when the MS is typeset, the product looks etched in stone. One cannot imagine how it could be any more perfect. Believe me, it can always stand improvement. You will have to retrain yourself to read your typeset work critically.

(For the flip side of the last paragraph, consider this. I was recently asked, by an important publishing house, to evaluate a manuscript for a textbook that they were considering developing. The manuscript was *handwritten*. This flies in the face of all that is holy; a manuscript that is going to a publisher should always be typed or word processed. In any event, I took what I was given and wrote my report. But this was a difficult process for me. I had to keep telling myself that this was *not* a rough set of notes, that it was a polished manuscript—even though it was handwritten and *looked* like a rough set of notes. Play this paragraph off against the last one for a lesson about form over substance.)

Once you become accustomed to writing with a computer, you will want to consider buying a "notebook computer". The notebook is a computer that is about as large as a statistics textbook, but has all the features of a desktop computer. The price of this tool has come down dramatically in the last few years, making it within reach of many. The advantages of a notebook computer are these:

1. You can take it with you when you travel.

2. You can use it when in the dentist's waiting room or when waiting for a plane flight.

3. You can take it with you to the library when you are preparing a bibliography.

4. On a nice day, you can sit under a tree outdoors and do your work.

5. With a cellular connection, you can even process your *e*-mail while on the run.

The notebook computer is one of those tools that, once you have it, you will wonder how you ever lived without. Of course your notebook computer can double as your desktop machine—either with a "docking station" or just by sitting on your desk. A notebook computer can drive a printer, can be hooked up to an external monitor or external keyboard, and can be equipped with a modem or a CD-ROM drive. When I am a visiting scholar at another university—for a week, a month, or a year—I usually take my notebook computer. My notebook is my home away from home. With an Iomega ZIP Drive® or a Syquest EZ Drive® I easily can carry all my data with me and/or transfer it to my notebook computer. My notebook simplifies my life.

And now a coda on backups. If you use a computer for your work, then develop the habit of doing regular backups. The "by the book" method for doing backups is the "modified Tower of Hanoi" protocol. This gives you access to any configuration that your hard disc has had for the past several weeks. Not all of us are up to that level of rigor. But do *something*. At least once per week, back up all your critical files to a floppy, a tape, a Bernoulli cartridge, a ZIP drive cartridge, or some other auxiliary device. Losing a hard disc is analogous to having your house burn down. It is an experience that you can well do without. Regular backups are a nearly perfect insurance against such a calamity.

6.2 Word Processors

I have already indicated in Section 6.1 the advantages that working on a computer has to offer. Next I shall specialize down to word processors and what they do. (I do this in part so that, when you read Section 6.5 about TEX, you will appreciate the differences.)

A word processor is a piece of software; you use this device for entering text on the computer screen, and for saving the text on a storage device (usually a disc). You engage in this process by striking keys on a keyboard—very similar to typing. The word processor performs many useful functions for you:

1. When you get to the end of a line, the word processor jumps to a new line—you do not have to listen for a bell, or keep one eye on the text, as you did in the days of typewriters.

2. The word processor allows you to insert or delete text, or to move blocks of text from one part of the document to another, with ease and convenience. You can create a new document (such as a letter) by making a few changes to an existing document.

3. The word processor right justifies (evens up the right margin) of your document. This process results in a more polished look.

4. The word processor can check your spelling.

5. The word processor communicates with your printer, and ensures that the document is printed out just as it appears on the screen.

6. The word processor enables you, if you wish, to incorporate graphics into your document.

7. The word processor allows you to perform "global search and replace" functions. For example, if you are writing a paper about mappings, and you decide to change the name of your mapping from F^* to G_k, then this can be done *throughout the paper* with a few keystrokes.

8. The word processor allows you to select from among several different fonts: roman, boldface, italic, typewriter-like, and so forth.

These days, most professional people prepare their documents on a word processor. Using a word processor saves time, money, and manpower. From the point of view of a mathematician, a word processor is not entirely satisfactory. The primary reason is that a word processor will not typeset mathematics in an acceptable fashion. A typical word processor can display *some* mathematics, but not in a form similar to what you would see in a high quality book. The word processor cannot treat complicated mathematical expressions: a commutative diagram, the quotient of a matrix by an integral, or a matrix with entries that are themselves matrices. Even for simple mathematical expressions, such as a character with both a superscript and a subscript, the output from a word processor is nowhere near the quality that one would see in a typeset book.

Outside of mathematics—in the *text*—word processors fall short in that they often do not *kern* the letters in words; many word processors use monospaced fonts, just like a typewriter. This fact means that the word processor does not perform the delicate spacing between letters—spacing that *depends* on which two letters are adjacent—that is standard in the typesetting process. The word processor does not offer the variety of fonts, in the necessary range of sizes, that is ordinarily used in typesetting. The word processor does not have sufficient power to adjust horizontal and vertical spacing on the page—both essential for the demands of quality page composition.

Put in other terms, a word processor is constrained by the fact that it is WYSIWYG ("what you see is what you get"). Even a Super VGA screen is about 100 pixels (dots) per inch, while high quality printing is 1200 dots or more per inch. Since a word processor prints *exactly* what appears on the screen, it can format with no more precision than what can be displayed on the screen. TeX, by contrast, is a markup language. It gives typesetting and formatting commands. It can position each character on the page within an accuracy of 10^{-6} inches.

Because a word processor is WYSIWYG, any file produced by a word processor will contain hidden formatting commands. One side effect of this simple fact is that if you cut out a piece of text from a word processor file and move it to another part of the file then it may not format properly. As an instance, suppose that you have a displayed quotation (such a display usually has text with wider margins and space above and below). Snip that out using standard commands for your word processor and drop it in elsewhere; it will not format correctly and you will waste a lot of time fixing it up. Because TEX is a markup language, it does not suffer this formatting malady. One of the beautiful features of TEX is that you can cut and move a fantastically complicated display and it will not change one iota.

A further point is that much mathematics is done these days on the Internet. A mathematician creates a document, sends it by *e*-mail to his collaborator in Australia, and awaits a revised version a few days later. A word-processed document contains hidden formatting commands—not just the ASCII code of characters and spaces that one obtains from a text editor (see Section 6.3). Such a document would not travel well on the Internet without extra help. You could UUENCODE the document, or send it in BINARY form. But then the recipient would have to know what to do to decode the document and get his hands on it. And, if not UUENCODE-ed, the document could easily be corrupted in transmission. TEX is much simpler, and much less prone to error.

Finally, no word processor is universal. There are too many word processing systems. They are all compatible to a degree, but not in the way that they treat mathematics. Thus, again, if one is doing mathematics using a word processor on the Internet then one will be hindered.

6.3 Using a Text Editor

Text editors are, primarily, for the use of programmers. A programmer wants an environment for entering computer code; the code will later be *compiled* by a FORTRAN compiler, a BASIC compiler, a C++ compiler, or some other compiler. Thus a text editor should not perform value-added features to the code that has been entered: there should be no formatting commands, no instructions for the printer, no hidden bytes, or any other secondary data. A file created with a text editor should comprise only the original ASCII code, together with space and line break commands.

A document printed directly from a file created with a text editor would look just like what you see on your computer screen—typewriter-like font and all—with a ragged right edge and with old-fashioned monospacing. Such a document might be acceptable for an in-house memo, but is not formatted in a manner that would be suitable for public use. Thus why would a mathematician want to use a text editor?

Today, TEX is the document creation utility of choice for mathematicians (see Section 6.5). Apart from its flexibility and the extremely high quality of its output, TEX is also infinitely portable and it is the one system that you can depend on most (and soon all) mathematicians knowing. If you want to work with a mathematician in Germany, using the Internet, and if you were to say to your collaborator "let's use Wordperfect®," then you would be laughed right off the stage. The only choice is TEX (or one of its variants, such as LATEX or AMS-TEX). And the point is this: TEX is a high level computing language (and also a *markup* language—see Section 6.5). You create a TEX document using a text editor.

Many a TEX system comes bundled with its own text editor. Usually, such a bundled editor has many useful features that make it particularly easy to create TEX documents. If you are a PC user, however, then you are accustomed to selecting your own software. The operating system DOS comes with a serviceable text editor called EDIT; many popular word processors, such as Wordperfect® and Wordstar®, have a "text editor" mode. But much more sophisticated text editors may be purchased commercially. Two of the best are Brief® (which is a version of the UNIX editor EMACS that has been adapted for the PC)[1] and Epsilon®. A good text editor can be customized for specific applications, allows you to open several documents and several windows at once, has sophisticated search and cut-and-paste operations, and will serve you as a useful tool.

---

[1] A Windows95® version of Brief® has recently been released under the name Crisp®. Epsilon® has a Windows95® version under development.

6.4 Spell-Checkers, Grammar Checkers, and the Like

The great thing about a document created on a computer is that the document is stored on your hard disc as a computer file. Thus your document has become a sequence of bytes. In most cases, your document in electronic form will consist primarily of ASCII code—ASCII is the international code for describing the characters that appear on your computer keyboard and your computer screen.

A computer file, consisting of a sequence of bytes, is grist for your computer's mill. The file is data ready to be manipulated. Apart from sending the data to a screen or to a printer, what else can your computer do with it? Here are some options:

- It can check the words in the file for spelling.

- It can check for repeated words, misused words, omitted words.

- It can check grammar and syntax.

- It can check style.

At this writing, spell-checkers are highly sophisticated tools. A good spell-checker can zip through a 10,000 word document of ordinary text in a minute or two. It will flag a word that it does not recognize, suggest alternatives, and ask you what you want to do about it. It will catch many standard typographical errors, such as typing "naet" for "neat," or such as typing the word "the" at the end of line n and also at the beginning of line $(n+1)$. Of course it will also flag most proper names, archaic spellings, and many foreign words and mathematical terms. As you use your spell-checker, you can augment its vocabulary (which is performed semiautomatically, so requires little labor), hence your spell-checker becomes more and more accustomed to *your particular writing.* Given that a spell-checker requires very little effort to learn and use, and that it can only add to the precision of your document (it suggests changes, and makes only those that you approve), you would be foolish not to use a spell-checker. *However:* Never allow the spell-checker to lull you into a false sense of security. To wit, the ultimate responsibility for correct spelling lies with you (see below for more on the limitations of spell-checkers).

If you use a garden variety spell-checker on a TeX document, then you will be most unhappy. The spell-checker will flag every TeX command (words beginning with \ ) and every math formula (set off by $ signs). You will find processing even a short TeX document to be an agony. Fortunately, the spell-checker Microspell® has a "TeX mode"; in that mode, Microspell® knows to ignore TeX commands and math formulas.

Do not use a spell-checker foolishly. If you intend to write the word "unclear" and instead write "ucnlear" (a common transposition error), then the spell-checker will certainly tell you, and that is useful information. But if you intend "unclear" and instead write "nuclear", then the spell-checker will forge blithely ahead—because "nuclear" is a *word*, and a spell-checker will only flag non-words. If you mean to say that someone is "weird" and instead you say he is "wired", then your message may still trickle through; but your spell-checker will not help you to get it right. The lesson is clear (rather than unclear): if your document passes the spell-checker, then you know that certain rudimentary errors are not present; however certain other, more sophisticated, errors could be present. Will you have to catch them yourself, with old-fashioned proofreading? Read on.

There also exist software tools based on artificial intelligence research, such as Grammatik®, that will critique your grammar, syntax, and usage (other similar software tools—available in the UNIX environment—are diction, double, explain, and style). These tools, while interesting, have a ways to go in sophistication. For example, Grammatik® gives knee-jerk criticism of all uses of the passive voice. It automatically praises all simple declarative sentences, whether they say anything (or contain standard English idiom) or not. It counts the words in your document and renders an opinion based on that single number. Clearly you must use a tool like this one with the proverbial grain of salt. It may help you to dig out obvious errors—split infinitives, mismatched subject and verb, etc.—but it is not the final arbiter of good English writing. Not every "error" that the spell-checker catches is truly an error (try including "pseudoconvex" in your manuscript), and not every "error" that Grammatik® catches is truly an error. Any software tool can pick out a non-error and call it an error. Couple the information that these software tools provide with a large dose of good sense. The final decision on how to use the information is yours.

An interesting DOS tool, available also in UNIX and other computer environments, is the "word counter". Run an ASCII file through the word counter and this utility reports **(i)** how many words there are in your document, **(ii)** what is the most frequently used word and how often the word is used, **(iii)** what is the second most frequently used word and how often is it used, etc. This device can easily be construed as an example of using a computer to do something just because it can be done. But we all have personal foibles, and the word counter can help to detect them. I, for example, tend to overuse the word "really"; I had to make special passes through this manuscript to weed out many occurrences of that word. A more informed opinion about which words you overuse can be made if you use word-counting software. If you use the word "really" more often than you use the word "the", then you may be in trouble; however, if you use it less often than you use the word "flagellate", then a different conclusion is in order.

I read of a professional author being stymied for a period of a year as a result of using word counting and other software. He ran one of his famous stories through the software, and it pointed out certain words and phrases that he overused. Thereafter, whenever the author was writing and began to use one of the pegged words or phrases, he panicked. The matter became worse and worse, and he eventually developed a writer's block. It took him considerable therapy, not to mention stress and hard work, to defeat this block.

We conclude with another anecdote, courtesy of G. B. Folland. One of Folland's publishers used a spell-checker that recognized the word "homomorphism" but not the word "homeomorphism". The result? The copy editor changed every instance of the latter to the former. The original manuscript contained several dozen of each. Now do you see how a spell-checker can get a person into trouble?

6.5   What Is TEX and Why Should You Use It?

TEX, created by Donald Knuth in the early 1980's, is an electronic typesetting system. Designed by a mathematician, specifically for the creation of mathematical documents, it also is a versatile tool in other typesetting tasks. In fact TEX is used in many law offices, and is also used to typeset *TV Guide*. The reference [Kn] tells something of the philosophy behind the creation of TEX.

What makes TEX such a powerful tool? First, TEX is almost infinitely portable. A TEX document created with a Macintosh computer in Hong Kong can be sent over *e*-mail to a PC user in Sheboygan who in turn can send it on to the user of a Cray I in Bielefeld. During this process, there are never any problems with compiling, printing, or viewing.

The book [SK] already contains this author's efforts at describing the inner workings of TEX and how to learn them. I shall not repeat that material here. Instead, I shall say just a few words about how TEX is used.

TEX is *not* a word processor. Instead TEX is what is called a "markup language". "Markup language" means that, in your TEX document (created with a text editor—not a word processor), you enter commands that tell TEX what you wish to have appear on each page, and in what position. TEX allows you to position each character on the page to within 10^{-6} inch accuracy.

If you think about all the material that appears on a typeset page, then what is described in the last paragraph sounds arduous—like it is simply too much trouble. Fortunately, TEX performs most typesetting tasks automatically.

If you are typesetting ordinary prose, then you simply type the words on the screen, with spaces between consecutive words. With TEX, you can leave any amount of space between successive words in your source code; you can also put any number of words on each line of code. TEX will choose the correct spacing, and the correct number of words for each line, when it compiles the document. You indicate a new paragraph by leaving a blank line. There is almost nothing more to say about typesetting text: TEX chooses the correct amount of space to put between words, how to put space between paragraphs, and so forth. It makes each line come out flush right, and ensures that each page contains the correct number of lines—not too many and not too few.

For mathematics, there are English-language-like commands that tell TEX just what you want. I will present just one example: The code

```
Now it is time to do some mathematics---a task for
which, given that we have spent many years at the
university, we are eminently well prepared.  Our work
is inspired by the identity $X(1 + X) = X + X^2$.


Let us consider the equation
$$
\int_X^{X^2 - X} {{\alpha^3
   + 17{{\alpha} \over {\alpha-2}}} \over
    {{\alpha-5} \over {\alpha + 1}}} \, d\alpha
     = \hbox{det} \, \left ( \matrix
    {X^2 & 3X & X \cr
  {{X^3 - 4} \over {X + 1}} & \sin X & \log X \cr
  {{X} \over {X+1}} & \hbox{erf}\, X & \sqrt{X} \cr
      } \right )
$$
which has been a matter of great interest in recent years.
```

would typeset as

Now it is time to do some mathematics—a task for which, given that we have spent many years at the university, we are eminently well prepared. Our work is inspired by the identity $X(1 + X) = X + X^2$.

Let us consider the equation

$$\int_X^{X^2-X} \frac{\alpha^3 + 17\frac{\alpha}{\alpha-2}}{\frac{\alpha-5}{\alpha+1}}\, d\alpha = \det \begin{pmatrix} X^2 & 3X & X \\ \frac{X^3-4}{X+1} & \sin X & \log X \\ \frac{X}{X+1} & \operatorname{erf} X & \sqrt{X} \end{pmatrix}$$

which has been a matter of great interest in recent years.

Even though you may not know TEX, you should have little difficulty seeing the correspondence between the code that is entered and the resulting output. (Note that the single dollar signs signify material to be typeset in "in-text" math mode; the double dollar signs tell TEX first to enter, and then to exit, "displayed" math mode.)

After you have created an `ASCII` file with your text editor, call it `MYFILE.TEX`, then you compile it with the command `TEX MYFILE` (variants are `LATEX MYFILE` and `AMSTEX MYFILE`). This creates the "device independent file", called `MYFILE.DVI`. The `DVI` file can be ported to a printer, to a screen, or translated to `Postscript`®.

As you can see from the preceding example, TEX does a magnificent job of typesetting mathematics. Usually no human intervention is required in order to obtain the quality and precision that you desire.

One interesting feature of TEX is that you cannot expect to see on the screen exactly what you will obtain in your printed output. For even a high quality super VGA screen has resolution less than 100 pixels per inch. Today, printers have a resolution of 600 or more dots per inch. (Note that 1200 dots per inch is production quality for many publishers.) The `Preview`® programs that come with TEX allow you to view your document to the extent of seeing where the various elements appear on the page—sufficient for doing elementary editing. But, to view the final output accurately, you must print a hard copy.

If you are using a `Postscript`® printer, you can use the command `\psfig` (or something similar) to import `Postscript`® graphics into your TEX document. Sophisticated systems like the NEXT® computer (or the `NEXT-Step`® operating system) allow you to cut and paste graphics and text in marvelous, and marvelously easy, ways.

TEX was originally designed with the notion of maximum power and flexibility in mind; Knuth planned that each discipline would develop its own style files to tailor TEX to its own uses. The variant LATEX endeavors to serve all end users. More specialized style files are available from the American Mathematical Society (to give just one example); these enable the AMS-TEX user to typeset a paper in the style of any of the AMS primary journals.

There is a whole new world of document-preparation tools available today. As a semi-neanderthal, I would be more than sympathetic if you do not want to dive into all the graphic and typesetting and electronic features that I have described here. In fact these tools are best learned in gradual stages. The learning curve for TeX alone is rather steep, although the book [SK] makes strides toward jump-starting the learning process. My recommendation is to begin by learning some form of TeX. LaTeX is a particularly popular form of TeX, and one favored by publishers (because it is more structured and steers the author toward more standard formatting styles than does Plain TeX). The reference [SK] creates an accessible bridge between Plain TeX (the most flexible TeX tool) and LaTeX (the least flexible TeX tool). Most mathematics departments have the hardware, the software, and the expertise to make it easy for you to learn TeX. This software is one of today's standard mathematical tools. You are shooting yourself in the foot not to learn it.

6.6 Other Document Preparation Systems

The material presented in the preceding sections suggests, rather forcefully, that TeX is the document-preparation system of choice in the modern mathematics environment. I shall not soft-pedal that message in the present section. However, there are various "front ends" for TeX that you may wish to consider.

`Publisher`®, by ArborText, is a graphics interface front end (that is `Postscript`® based) for TeX that you use by pointing at icons with a mouse. You see the compiled, typeset, version of your document *while you are writing* (this protocol is termed `WYSIWYG`). With `Publisher`®, you no longer need to learn TeX commands. For the Greek alphabet, you point at the Greek alphabet icon and choose the letter that you want. For a formatting command, you point at the formatting icon and choose the command that you need. Users of `Publisher`® rarely, if ever, refer to the documentation. The downside of `Publisher`® is that it requires a rather sophisticated hardware platform. Also the output from `Publisher`® is not exactly TeX code, but something that approximates TeX code. By the time this book is in print, your desktop computer may be adequate for `Publisher`®, and `Publisher`® output may have become standardized.

`Scientific Workplace`®, by TCI, is a graphics interface front end for LaTeX that works on an ordinary PC of level 386 or higher. This product is also `WYSIWYG`. It works with pull-down menus and some icons. One of its most powerful features is that it interfaces with `Maple`®. For instance, if you typeset an integral and follow it with an equal sign, then the software will evaluate the integral for you automatically—*inside your document.* Imagine what a boon this tool is to textbook writers!

The Lightning version of `Textures`® for the Macintosh computer, created by Blue Sky Publishing, works with a split screen. On the left screen your TeX code is displayed as you enter it. On the right screen your document is compiled on the fly and displayed `WYSIWYG`. Lightning is a popular working environment.

I must say, without naming names, that all the systems named above have bugs and shortcomings. But they are improving all the time, and when they work they are beautiful. Any of the systems described here can help an otherwise timid neophyte over the hump and into the world of computer typesetting. They are powerful and useful tools.

6.7 Graphics

As indicated elsewhere, the most common method for including graphics in a book is still to create them *separately*, each on its own page. The drawings could be created by hand, with pen and ink. Or they could be produced with Corel `DRAW!`®, or `MacDraw`®, or Harvard `Graphics`®, or any number of other packages. To repeat, each figure should be on a separate piece of high quality drawing paper (available from any store that carries art supplies) and drawn in dense black ink. Use a proper drawing pen—not a ball point, or a rolling writer, or a pencil. Best is to draw the figures (considerably) larger than they will actually appear in the book, in thick dark strokes. When they are photographically reduced to fit, then the pen strokes come out sharper, denser, and darker.

Each figure should be labeled clearly: a typical label might be

Chapter 3   Section 2   Figure 5

Correspondingly, somewhere in Section 2 of Chapter 3 there should be a space set aside for this figure, and it should be labeled "Figure 5". (I am assuming here, for simplicity, that you are producing your document in some version of TEX; if not, then forget about leaving a space in the document, but *do* put a label in the margin.) And be sure that the text contains a specific reference to each figure; do not leave it to the reader to determine what figure goes with which set of ideas. (The same remark applies, of course, to tables.) It helps, though it is not mandatory, to give each figure a title and a caption.

Drawing good illustrations for your work is an art. A good figure is not too busy, does not have extraneous information or extraneous pen-strokes, and displays its message prominently and clearly. The books [Tuf1] and [Tuf2] by Edward Tufte will give you a number of useful pointers on how to develop powerful graphics for your work.

Today, publishers are perfectly happy to "compose" each page of your work. In other words, the publisher will create a final hard copy, or "repro copy", with the art properly situated on each page. In point of fact, most publishers prefer to retain some control over the final appearance of the page. There are strict rules in the publishing world as to how art is supposed to appear (especially regarding space above and below a figure, and how the figure is to be centered left-to-right); most likely you do not know these rules, and you had best let the publisher handle the matter.

Of course we all know that there are copious electronic tools for creating artwork in your manuscript. Just as an example, many versions of TEX have simple commands, such as \psfig, that allow you to import an encapsulated Postscript® file into your document. In one common scenario, a \special command insets raw printer commands into the file that will communicate with your printer. The result is that your Postscript® figure appears right on the printed page (*provided* that you have a Postscript® printer or know how to use Ghostscript to make Postscript® talk to a non-Postscript® printer). Some versions of TEX—such as Personal TEX®—understand several other graphics languages as well. For example, the Hewlett-Packard language PCL is a graphics protocol designed for use with certain HP printers. And many graphics programs give you a choice of several different graphics output languages; these could include PS, EPS,

BMP, or `WMF` graphics images. The documentation for your T_EX software (for instance `Personal` T_EX®) will explain precisely which graphics languages it can handle and how it does so.

But now I shall contribute a few prejudicial remarks about the contents of the last paragraph. I write numerous books, many of which contain a great many figures, and I have never considered producing my figures electronically. And no publisher has ever asked me to do so. I have already said why publishers do not request it, and now I will say a word about why I do not do it.

The short answer is that I value my time. In most instances, I can produce a serviceable figure by hand, using drafting equipment, in a few minutes. Either the publisher uses my figure, or uses its in-house technical artists to produce the final figure (using my figure for guidance). A common scenario these days is that the technical artist scans in my figure and then tweaks it a bit using `MacDraw`® or Aldus `Pagemaker`® or Adobe `Illustrator`®.

Drawing a figure with software is usually slower, and requires a lot of fiddling. Positioning the figure on the page also requires a lot of fiddling. I just do not want to do it. Computer drawing has its place—for instance Computer Aided Design systems (CAD's) are a small miracle—but in my experience it often has no role in producing illustrations for a mathematical monograph.

No matter how your publisher handles your figures, always insist on seeing proof sheets of the final version of each figure. Mathematics has little tolerance for error: if your figure is supposed to exhibit certain lines intersecting in a certain way, or a tangency, or the congruence of certain angles, then it had better exhibit just that. The artist in the publisher's production department does not usually have the knowledge and expertise to know exactly what your figure is supposed to show. It is up to you, the author, to make sure the figure is correct.

Let me conclude with two *caveats*. First, some computer systems make the production of graphics relatively easy. As an example, the NEXT® computer has an integrated environment that allows you to drop your `Postscript`® graphic into your document by cutting and pasting with a mouse. You need not worry about making your different pieces of software talk to each other, about hardware and software compatibility, or about anything else. It just works. Unfortunately, the original NEXT® computers are no longer available. But the operating system `NEXT-Step`® makes a PC computer behave like an old black-box NEXT® computer, so the attractive working environment that I have just described is still available—and is notably faster on a PC than on an original black-box NEXT®.

My second *caveat* concerns `Mathematica`®, `Maple`®, and the like. These, too, are small miracles. If you need to draw a hyperboloid of one sheet, or the graph of $z = \log(|\sin(x^2 + y^3)|)$, then there is nothing to beat `Mathematica`®. I recommend that you use it. `Mathematica`® will output your figure in encapsulated `Postscript`®, for storage on your hard disc, and in principle this file can be imported into your document—so that it appears in a space that you have set aside for it in the text. My recommendation is that instead you should just print your `Mathematica`®-generated `Postscript`® file on a separate sheet of paper, as discussed above. Let the publisher (with a staff of professionals who know what they want) compose the pages of your book.

A final note: ask `Mathematica`® to graph a horrendously complicated function of two variables, and it will do so in an instant. Such tasks are what `Mathematica`® is designed to perform. And it will provide the labels on the axes automatically. But endeavor to draw a rectangle or triangle, and to label the vertices in your own fashion, and it may take you an hour. Conversely, I can hand draw the rectangle or triangle and provide the labels in five minutes. But it could take me hours to graph the function. *Use the proper tools in the proper context.*

6.8   The Internet and `hypertext`

Just a word about `hypertext`, and about electronic publishing in general. The spirit of electronic publishing is to bypass the traditional hard copy of published materials, and instead make the materials available on the Internet and the World Wide Web. Readers would be identified and would pay either by buying a password or by paying the publisher to make materials available to a *particular* CPU with a particular identification number (the IP address—given by four octets of code).

A part of this new electronic publishing environment is `hypertext`. With `hypertext`, certain words or phrases in the electronic document appear in an accented form—often in a different color or underlined. If the reader "clicks" on the accented word, then he is "jumped" to a cognate item. For instance, if you are reading a book on the function theory of several complex variables, you come across the word "pseudo-convex", and you cannot recall what it means, then—instead of madly flipping through the book trying to find the definition (that is the old way)—you click on the word and are jumped either to the passage that contains the definition, or perhaps to a lexicon, or perhaps to a menu that offers you several options. Alternatively, you could click on a reference to another book or paper and you would be jumped to the reference—to the *actual text of the reference*—no matter where in the world the source is. Or you could click on the name of a mathematician who is mentioned in the text and you would be jumped to his home page, or to his publication list. Yet another scenario is that you could click on an icon or a button and bring up animated graphics.

Clearly `hypertext` is an amazing device, and the possibilities that it offers are vast and powerful. In the coming years, as the world decides what role electronic publishing will have in our lives, how to charge for it, how to market it, how to archive electronic documents, and so forth, we will see more and more electronic books and documents. For now, matters are in a developmental stage.

There now exist many electronic journals. An electronic journal is one in which all transactions—submission, remanding to a referee, referee's report, editorial decision, and publication—are executed over the Internet.

Several of these new electronic journals are "startup" journals, run "for love" by an individual from his office computer. Others are institutionalized, but are still free.

At the beginning of 1995, the AMS (American Mathematical Society) eliminated the "Research Announcements" section from the *Bulletin* of the AMS and is instead publishing research announcements—with essentially the same editorial policies and publishing standards—in electronic form. The AMS is also making archival/disaster-backup copies of all the startup electronic journals, strictly as a service to the mathematical community and at no charge. As of this writing, the

AMS has initiated several brand new subject-area electronic journals for which subscribers will pay a modest fee.

Most electronic journals are run with the same editorial procedures as for a paper journal (these procedures are described in Section 2.7). The primary differences are two:

1. With an electronic journal, page limitations are not important; therefore an electronic journal can publish longer articles and more of them.

2. The avowed goal of most electronic journals is to generate no paper—none whatsoever. Therefore papers are submitted by *e*-mail and forwarded to the editor and to the referee by *e*-mail. The referee's report is sent back to the editor, and then to the author, by *e*-mail; and any revisions are resubmitted by *e*-mail. The paper is published electronically, either on a bulletin board or on a server. There is never a hard copy of anything. The published paper is posted in an output language, such as *.DVI (a Device Independent file generated by TeX) or *.EPS (an encapsulated Postscript® file) or *.PDF (an Acrobat® file). *There are no reprints*, and there are no hard copies of the journal. In some cases, CD-ROM versions of the electronic journal are available for archival purposes. Usually the end user can download individual papers for (compiling and) hard copy printing for personal use.

Some hard copy journals are now simultaneously publishing an electronic version. One interesting innovation is that some traditional journals make any mathematical paper available electronically, for a modest charge, as soon as the paper has been accepted.

The world of electronic publishing is just opening up, and promises new frontiers of publishing activity and also of legal complications. As an example, the copyright law issues connected with electronic publishing are immense [Oke].

Some authors are making entire books available at no charge on the World Wide Web. Commercial publishers are also exploring the publication of electronic forms of books. In fact some publishers will propose to an author that a home page be set up for his book, and that not simply the book but also a variety of ancillaries appear on the web site. These ancillaries could include relevant papers, a bibliographic database, exercise books, lecture manuals—you name it. In some scenarios, a publisher may develop a version of a book to which readers may contribute interactively. As of this writing, the commercial electronic publishing of books is nascent.

6.9   Collaboration by *e*-Mail; Uploading and Downloading

Writing a collaborative mathematical work is a source of great pleasure. It is especially fun when you use *e*-mail as a tool. Entire chapters can be zapped around the world in an instant. You get immediate feedback on your ideas. In many (but not all) ways, collaborating by *e*-mail is like having your partner in the office next door.

Many of us do our work on the computer system at school. The school system is probably a network—most likely a UNIX system. If this describes your working environment, then Internet collaborative work is a breeze. You work on your document—using an editor like VI or EMACS. When you are ready to share it (call the paper OURPAPER.TEX) with your collaborator, you type

```
MAIL SK@MATH.WUSTL.EDU < OURPAPER.TEX
```

and press <Enter>.

Here is a translation of what you just saw. I am assuming that your collaborator is me. My *e*-mail address is SK@MATH.WUSTL.EDU. That explains the first part of the displayed command. The character < is an "include" sign. What is being included is the TeX file OURPAPER.TEX. Once you press <Enter>, your paper OURPAPER.TEX will be automatically zapped off into the ether (and will land in my mail spooler). Of course you will still retain the master copy on your system's hard disc. (Note that in Section 4.6 a method is described for "including" a document in an *e*-mail message that you are writing. This provides an alternative method for sending your document over the Internet.)

If you wish, you can type comments at the beginning of, or in the middle of, the TeX document. If you precede each line of the comment material with a % symbol, then TeX will ignore those lines.

Perhaps you do your work at home, on a PC or a Macintosh, and then bring the files to work for further processing and *e*-mailing. Thus you copy the files to a diskette and must "upload" the files from the diskette to the computer system at school. You will have to consult your local guru for the details of this uploading process. But note this: there are differences in the ways that files are formatted on a microcomputer as compared with a mini or mainframe computer that uses the UNIX operating system.

One of the most significant of these differences is that line breaks in files that are created in the DOS or Macintosh (MAC) operating systems are different from linebreaks in UNIX. If you upload a file from a DOS or Macintosh diskette to UNIX (a SPARC® station has a diskette drive that can read a 3.5-inch disc, so this task is easy to perform) and immediately attempt to edit it (using EMACS, for example) then you will find that the file has either no line breaks or garbled line breaks. A similar thing will happen if you "download" a file from the UNIX operating system to a DOS or a Macinosh system. Thus you must use the utilities DOS2UNIX or UNIX2DOS (for translating from DOS to UNIX or UNIX to DOS, respectively) or MAC2UNIX or UNIX2MAC (for translating from MAC to UNIX or UNIX to MAC, respectively) to change the line breaks before you attempt to work with the file on the new system.

The public domain operating system LINUX for the personal computer is a popular choice these days. LINUX is a version of UNIX that is designed for PC-type computers (there is also a version of UNIX designed for Macintosh computers). In LINUX you can open either a DOS window or a Macintosh window, and the different operating systems can talk to each other. If your personal computer runs on LINUX, then you can cut through a lot of the red tape that was described in the last paragraph. You can purchase a "LINUX Starter Kit", including a CD-ROM with the operating system and a book that documents the software, for under $30.

Chapter 7
Closing Thoughts

Of all those arts in which the wise excel,
Nature's chief masterpiece is writing well.

John Sheffield, Duke of Buckingham and Normanby
Essay on Poetry [1682]

Great prizes are reader interest and understanding; all else is secondary. Graceful prose, imagery, wit, even orthography and grammar are only means to more important ends. This observation makes writing and reading more of a colloquy and less a lonely or isolating business.

from the dust jacket of *Mathematical Writing* [KnLR]

England has forty-two religions and two sauces.

Voltaire

A writer needs three things, experience, observation, and imagination, any two of which, at times any one of which, can supply the lack of the others.

William Faulkner

Isaac Newton invented his theory of gravity when he was 21. I'm 32, and I just found out that Garfield and Heathcliffe are two different cats.

Anon.

Anything that helps communication is good. Anything that hurts is bad.

Paul Halmos

7.1   Why Is Writing Important?

The case for writing, indeed for good writing, has been made throughout this book. Writing is our tool for communicating our ideas, and for leaving a legacy for future generations. One of the marvels of genuinely outstanding writing is its longevity. In many ways, the writings of Herodotus, of Descartes, of Plato, or of Faulkner are as vibrant and important today as when they were first penned.

Writing at the very highest level is often painstaking and tedious. Good authors can spend an entire day agonizing over a word, a comma, or a phrase. They revise their work mercilessly. For the working mathematician, I am not recommending this sort of writing. Do it if you like; but this level of precision and artistry is not what our profession either demands or needs. In fact the sort of clear, cogent, precise writing that I am promoting here requires little more effort than lousy writing requires. Like the ability to scuba dive, the ability to write well is in truth a matter of becoming conversant with the basic principles and then practicing. Once you become comfortable with the process, then writing becomes less of a chore and more of a pleasurable pastime. It allows you to view your written work as an accomplishment to be proud of, rather than another agony that you have slogged your way through.

We all grow up speaking English (or some other native language). After a while, we convince ourselves that we are able to express our thoughts verbally—regardless of our technical facility with grammar, usage, and syntax. As we grow older, a corollary of such reasoning is that we all think that we know how to write. A result of this process is that it is more difficult to teach people to write (and, in turn, for them to learn to write) than it is to teach people calculus. When a student has his calculus work marked incorrect, then he is inclined to say "Apparently I don't know how to do this kind of problem. I'd better get some help." But when a student has his writing marked up and criticized, then he is liable to go to the instructor and say "Well, just what is it that you want?"

Learning to write well is a yoga; it is a manner of being trained in self-criticism and self-instruction. Fields Medalist Enrico Bombieri has observed to me that his artistic activities, particularly his painting, have helped him to see things more clearly, and in greater detail. Just so, learning to write well will sharpen your thoughts, develop your skills at ratiocination, and help you to communicate more effectively.

Developing an ability to write effectively will give you an appreciation of the writing, and the thinking, of others. And you will learn from their writing—both what to do and what not to do. It will add a new dimension to your life. I hope that it is a happy one.

Bibliography

[Ad1]   *The Adams Cover Letter Almanac and Disc*, Adams Media Corporation, Holbrook, Massachusetts, 1996.

[Ad2]   *The Adams Jobs Almanac*, Adams Media Corporation, Holbrook, Massachusetts, 1996.

[Ad3]   *The Adams Resumés Almanac and Disc*, Adams Media Corporation, Holbrook, Massachusetts, 1996.

[BM]   G. Birkhoff and S. MacLane, *A Survey of Modern Algebra*, MacMillan, New York, 1941.

[Blo]   S. Bloch, Review of *Étale Cohomology* by J. S. Milne, *Bulletin* of the AMS, new series, 4(1981), 235-239.

[Chi]   *A Manual of Style*, 14th Edition, (Chicago: The University of Chicago Press, 1993).

[Dub]   E. Dubinsky, A. Schoenfeld, J. Kaput, eds. and T. Dick (managing ed.), *Research in Collegiate Mathematics Education, I*, The AMS in cooperation with the MAA, Providence, Rhode Island, 1994.

[DS]   N. Dunford and J. Schwartz, *Linear Operators*, Wiley Interscience, New York, 1958-1971, 1988.

[Dup]   L. Dupré, *Bugs in Writing*, Addison-Wesley, Reading, Massachusetts, 1995.

[Dys]   F. Dyson, Missed Opportunities, *Bulletin* of the AMS 78(1972), 635-652.

211

[Fow] H. W. Fowler, *Modern English Usage*, Oxford University Press, Oxford, 1965.

[Fra] M. Frank, *Modern English: A Practical Reference Guide*, 2nd edition, Regents/Prentice-Hall, Englewood Cliffs, New Jersey, 1993.

[Gil] L. Gillman, *Writing Mathematics Well: A Manual for Authors*, The Mathematical Association of America, Washington, D.C., 1987.

[Gle] J. Gleick, *Chaos: Making a new Science*, Viking, New York, 1987.

[GH] P. Griffiths and J. Harris, *Principles of Algebraic Geometry*, John Wiley and Sons, New York, 1978.

[Hig] N. J. Higham, *Handbook of Writing for the Mathematical Sciences*, SIAM, Philadelphia, Pennsylvania, 1993.

[Hor] J. Horgan, The Death of Proof, *Scientific American*, October, 1993.

[JQ] A. Jaffe and F. Quinn, "Theoretical Mathematics": Toward a cultural synthesis of mathematics and theoretical physics, *Bulletin* of the AMS 29(1993), 1-13.

[Kli] W. Klingenberg, Review of *Affine Differential Geometry* by K. Nomizu and T. Sasaki, *Bulletin* of the AMS, new series, 33(1996), 75-76.

[Kn] D. E. Knuth, Mathematical typography, *Bulletin* of the AMS (new series) 1(1979), 337-372.

[KnLR] D. E. Knuth, T. Larrabee, and P. M. Roberts, *Mathematical Writing*, Mathematical Notes No. 14, The Mathematical Association of America, Washington, D.C., 1989.

[Kr1] S. G. Krantz, *Real Analysis and Foundations*, CRC Press, Boca Raton, Florida, 1991.

[Kr2] S. G. Krantz, *How to Teach Mathematics*, American Mathematical Society, Providence, Rhode Island, 1992.

[Kur] K. Kuratowski, *Introduction to Set Theory and Topology*, Addison-Wesley, Reading, Massachusetts, 1961.

[Lam] L. Lamport, LaTeX*: A Document Preparation System. User's Guide and Manual.*, 2nd Ed., Addison-Wesley, Reading, Massachusetts, 1994.

[Lan] S. Lang, Mordell's review, Siegel's letter to Mordell, Diophantine geometry, and 20th century mathematics, *Notices* of the AMS 42(1995), 339-350.

[Lip] J. Lipman, Review of *Principles of Algebraic Geometry* by P. Griffiths and J. Harris, *Bulletin* of the AMS, new series, 2(1980), 197-200.

[Man] B. Mandelbrot, *Bulletin* of the AMS 30(1994), 193-196.

[McL1] M. McLuhan and Q. Fiore, *The Medium is the Massage*, Random House, New York, 1967.

[McL2] M. McLuhan and B. Powers, *The Global Village*, Oxford University Press, New York, 1989.

[MW] *Merriam-Webster's Dictionary of English Usage*, Merriam-Webster, Inc., Springfield, Massachusetts, 1994.

[Mit] M. Mitchell, *Gone with the Wind*, Macmillan, New York, 1936.

[Moo] G. E. Moore, *Ethics*, Holt and Company, New York, 1912.

[Mor] L. Mordell, Review of *Diophantine Geometry* by S. Lang, *Bulletin* of the AMS 70(1962), 491-498.

[New] E. Newman, *Strictly Speaking*, Warner Books, New York, 1974.

[NoS]   K. Nomizu and T. Sasaki, Book Review by Klingenberg, *Notices* of the AMS 43(1996), 655–666.

[Oke]   A. Okerson, Whose article is it anyway?, *Notices* of the AMS 43(1996), 8-12.

[PS]   H. O. Peitgen and D. Saupe, *The Science of Fractal Images*, Springer-Verlag, Berlin, 1989.

[Por]   R. Porter, *Mathematics into Type*, revised edition, AMS, Providence, Rhode Island, in preparation.

[Saf]   W. Safire, *New York Times*, January 8, 1989, Section 6, p. 12.

[SK]   S. Sawyer and S. G. Krantz, *A TEX Primer for Scientists*, CRC Press, Boca Raton, Florida, 1995.

[SG]   M. E. Skillin, R. M. Gay, et al, *Words into Type*, Prentice-Hall, Englewood Cliffs, New Jersey, 1994.

[Ste]   N. Steenrod, et al., *How to Write Mathematics*, the AMS, Providence, Rhode Island, 1973.

[SW]   W. Strunk and W. White, *The Elements of Style*, 3rd Edition, Macmillan, New York, 1979.

[Swa]   E. Swanson, *Mathematics into Type*, revised edition, the AMS, Providence, Rhode Island, 1979.

[Tuf1]   E. Tufte, *The Visual Display of Quantitative Information*, Graphics Press, Cheshire, Connecticut, 1983.

[Tuf2]   E. Tufte, *Envisioning Information*, Graphics Press, Cheshire, Connecticut, 1990.

[VanL]   M. C. van Leunen, *Handbook for Scholars*, revised ed., Oxford University Press, New York, 1992.

[Wer]   J. Wermer, Letter to the Editor of the *Notices* of the AMS 42(1995), 5.

[WR]   A. N. Whitehead and B. Russell, *Principia Mathematica*, Second Edition, Cambridge Univ. Press, Cambridge, 1950.

Index